METHODS IN RESEARCH AND DEVELOPMENT OF BIOMEDICAL DEVICES

METHODS IN RESEARCH AND DEVELOPMENT OF BIOMEDICAL DEVICES

Kelvin K L Wong
Jiyuan Tu
Zhonghua Sun
Don W Dissanayake

World Scientific

NEW JERSEY • LONDON • SINGAPORE • BEIJING • SHANGHAI • HONG KONG • TAIPEI • CHENNAI

Published by

World Scientific Publishing Co. Pte. Ltd.
5 Toh Tuck Link, Singapore 596224
USA office: 27 Warren Street, Suite 401-402, Hackensack, NJ 07601
UK office: 57 Shelton Street, Covent Garden, London WC2H 9HE

British Library Cataloguing-in-Publication Data
A catalogue record for this book is available from the British Library.

METHODS IN RESEARCH AND DEVELOPMENT OF BIOMEDICAL DEVICES

ISBN 978-981-4434-99-7

Printed in Singapore by World Scientific Printers.

Foreword

Innovation is the lifeblood of medical device manufacturers. What constitutes innovation? This is what this book is all about — developing a system engineering methodology for the research and development of medical devices. The book concerns the modeling, design, testing, and development of medical (and particularly biomechanical) devices, namely prosthetic heart valves, endovascular stents, biomedical micro electromechanical systems (Bio-MEMS) micropump, and nasal drug delivery spray devices.

The modeling, design, and testing constitute the most important elements of prototyping a medical device. After design, prototyping, and testing, the device goes into the manufacturing and validation process, which is even more demanding. The major components of the product development cycle include concept, development, testing, regulatory approval, production, and ultimately, marketing. However, before the device enters the manufacturing stage, it is important that the end users (namely the physicians and particularly the surgeons) are intrinsically involved in the research and development process.

The physician and surgeon define how the device is going to be deployed in order to carry out its function. Their involvement is hence critical in helping to refine the final design features of the device, the instrumentation for deploying the device, the surgical technique for the implantable device, and the delivery method. For instance, in the case of prosthetic valve design and stent design, the implantation methodology needs to be part and parcel of the device design and development.

Consequently, the engineering research and development team needs to take the responsibility for incorporating the vision of the physician and surgeon champions to create a device that is cost-effective, easy to use, easy to manufacture, safe, and effective, while employing cutting-edge technologies. After the design is complete, a comprehensive assessment including failure modes and effects analysis, needs to be performed. This process will provide understanding of the potential risks and liabilities to a company in employing the medical device.

Before the start of the design process, it is necessary to know the anatomical and physiological environment in which the implant device (such as the prosthetic aortic valve and the coronary stent) and the drug delivery device are to be working. Medical imaging such as magnetic resonance imaging (MRI) and computed tomography (CT) can produce the requisite images of anatomical or functional features *in vivo*. These images can be processed and reconstructed by using computer modeling tools, to provide the requisite visualization of the anatomical scenario of the medical device's operational environment.

This is followed by device design by using computational modeling. For instance, in the case of the *prosthetic aortic leaflet valve*, there is a need to first prepare the structural model of the valve. Both finite element design of the valve and the computational fluid dynamics (CFD) of blood flow into and out of it need to be analyzed in an iterative fashion until the design effectively demonstrates how the blood moves into and is forced out of the leaflet pockets. After the computational modeling process, the device is tested *in vitro* and *in vivo*. The *in vitro* measurements are based on a mechanically confined environment with the help of measurement sensors. Herein, flow visualization technology packages are employed for the analysis of flows in the space within and distal to the heart valve. Thereafter, *in vivo* measurements of flows through and distal to the valve provide a more realistic perspective of how the valve device will serve its anatomical and physiological functions.

Coronary endovascular stenting was introduced into clinical practice for myocardial revascularization to treat blocked coronary vessels, as a less invasive technique relative to coronary arterial bypass grafting (CABG). While stent and stent grafts are increasingly used in clinical practice as an effective alternative to open surgery, the long-term outcomes are yet not comparable to CABG. One of the main concerns of implanting such devices into the blood vessels is thromboembolism or restenosis, due to the interference of regional blood flow by the implanted devices. Hence, further studies are needed to enhance the safety of these coronary stents, by using a combination of three-dimensional (3D) visualization and CFD. CT imaging with the aid of virtual intravascular endoscopy (VIE) can generate the 3D anatomical reconstruction of coronary vessels and inserted stents, to evaluate how the stent grafts situate within the occluded coronary vessel. The stent design then requires fluid–structure interaction (FSI) analysis to determine the flow within the stent as well as between the stent and the coronary vessel wall.

With regards to *nasal drug delivery device* design, the nasal airway is dominated by the nasal turbinates that are lined with highly vascularized mucosa that contain openings to the paranasal sinuses. Due to these characteristics, drug delivery to combat lung diseases, cancers, diabetes, sinus infections, etc. can be made viable by having the drug deposited in the turbinate region (instead of in the anterior regions of the nasal vestibule). Hence, CFD studies into nasal drug delivery can be carried out by constructing a 3D computational model of the nasal cavity from CT scans, with the sprayed drug particles delivered from the nasal spray device. After the reconstruction of the respiratory airways via medical imaging, CFD techniques can simulate the airflow patterns, particle trajectories, and the drug–particle interactions with the inhaled air that leads to particle deposition and location. Such results can provide a deeper insight into nasal drug delivery for the optimization of the spray device design. Now, recent advances in nanotechnology have enabled the manufacture of engineered nanoparticles leading to greater biologic activity desirable for drug delivery. It has been found that deposition patterns of 1-nm particles under a flow rate of 10 L/min in a human nasal cavity display an even distribution throughout the geometry, compared to deposition patterns for micron-sized particles being concentrated in the anterior third of the cavity.

Finally, we address the *Bio-MEMS micropump* device for micro-drug delivery applications. The number of patients suffering from life-threatening cardiovascular diseases (hypertension, heart attack, or stroke), septicemia, cancer, diabetes, melancholia, and malignant lymphoma has increased tremendously. Conventional drug delivery methods such as oral tablets or injections have various limitations. Furthermore, conventional controlled release formulations are designed to deliver drugs at a predetermined and preferably constant rate. Some clinical situations necessitate either an external control of the drug delivery rate or a certain volume of drug that is beyond the capabilities of existing controlled release formulations. Therefore, conventional methods are not effective in delivering a drug within their therapeutic range. A new alternative transcutaneous energy transmission system is required for microdevices targeted at *in vivo* applications. For these devices, effective techniques are required for accurate automated dosing capabilities, so that the patients are prevented from sudden death or the incorrect intake of the drug. Therefore, the implementation of targeted micro-drug delivery methods is reckoned to be a critical solution in lifesaving health advances.

In this context, a typical Bio-MEMS-based micropump constitutes an implantable drug delivery system, which provides the actuation source to effectively dispense drugs or therapeutic agents to a targeted area in the body with precision, accuracy, and reliability. It depends on the operation of a microdiaphragm that moves up and down to push biofluids through a channel. A radiofrequency (RF)-controllable microvalve or a micropump is deemed to be the next generation of intelligent implantable devices. Micropumps built on the concept of wireless transcutaneous RF communication in biomedical applications can generate a remotely controlled device for the regulation of drug fluid delivery.

The authors Dr. Kelvin K.L. Wong, Profs. Jiyuan Tu and Zhonghua Sun, and Dr. Don W. Dissanayake are to be congratulated for preparing this outstanding and insightful book, which presents a chart for research, design realization, testing and design analysis, marketing and production, as well as future development of advanced biomedical devices. Its contents include methodologies for design and product engineering, medical imaging, CFD, finite element analysis, computer-aided design, and rapid prototyping. This book can be a reference for a course on biomedical devices and can serve as a useful manual guide to engineers, medical doctors, radiologists, physicians, and surgeons that are involved in the research of medical technologies.

Dhanjoo N. Ghista
Professor: Engineering and Medical Sciences

Preface

In the twenty-first century, the public perception that health care has received international attention and a significant amount of financial support from the government is on the rise everyday. A main contributor of this phenomenon is believed to be the unrestrained utilisation of state-of-the-art technology that appears to lead to a potential remedy for a medical problem. However, such technologies are often overly expensive due to the high costs involved in research and development of such biomedical devices. Fortunately, with the recent advances in computing technology and manufacturing, the level of expenses for biomedical device development can be significantly reduced.

The development of a biomedical device or a medical-assisting mechanism involves the system integration of various modelling, evaluation or testing components that lead to the successful examination of a product that meets the medical application safety threshold. The nature of a device application can be diverse; it can be classified as a surgical correction mechanism, an introduction of a mechanical structure to rectify a defective function or to replace a failed part in the anatomy, a mechanism to deliver processes or drugs into the body or a computational device to aid a specific medical operation. The management of evaluating a medical application is fundamentally the execution of product modelling, prototyping, testing and remodelling procedures. Some of these procedures have evolved into current state-of-the-art system engineering processes that often involve other various instruments leading to a specific purpose or are deliverable. Research and development of medical technologies takes the form of modelling and evaluation of heart valve implants, stents, drug delivery mechanism, etc. It develops the instruments used in testing these cardiac or respiratory flow related biomedical applications, which enhance the human quality of life and increase the human lifespan. The system of analysing the testing and development of biomedical devices exists to facilitate a smooth flow of implementation that evaluates medical implants and mechanism in order to meet a desired level of reliability and clinical/surgical safety.

This book presents a roadmap for applying the stages in conceptualisation, evaluation and testing of new medical technologies in a systematic order of approach, leading to solutions for medical problems with a well-deserved safety limit. The contents we have prepared will pave the way for understanding the preliminary concepts used in modern biomedical device engineering, which includes medical imaging, computational fluid dynamics, finite element analysis, computer-aided design and rapid prototyping. They have been designed to give you information on the advances in biomedical technologies, and enable you to capitalise on the research and development methods present in this book for designing effective biomedical devices. We hope that this book can be of use to biomedical engineers, medical doctors, radiologists and any other professionals related to the research and development of medical devices for health care. It is also an ideal postgraduate teaching material and may also be promoted by lecturers as a textbook and manual guide.

Our material is thoroughly strategised and organised. We believe that its entrance into the scientific community will be highly beneficial and educational. We also expect interests by the medical defence and health industries. The book also contributes the framework behind the implementation of biomedical device and application in the systematic format, and describes the key instruments used in its development. Experimental and clinical verification will usually complement the existing computational results based on the device performance and vice versa. The understanding and analysis of flow properties during the device operation are necessary to evaluate its performance and safety. These concepts are presented by our book during the outline of the management in engineering and production for specific medical technologies. We consolidate the experimental and developmental works for a few biomedical flow devices, and organise them into concepts and case studies that describe the fundamentals of prototyping and evaluation. We examine the utilisation of these device evaluation methods on specific biomedical devices as exemplifications.

This book is organised as follows. In Chapter 1, the overall integration of approaches is reflected in a system diagram for developing and evaluating biomedical devices, and we shall provide the key descriptions of supporting instruments that assist in achieving the final product. Chapter 2 reviews the existing biomedical device and gives a preliminary background of the testing or evaluation methods used to enable the meeting of a safety threshold after implantation. Chapter 3 gives a detailed description of the computational modelling methods for the biomedical device, and outlines the related tools that can be utilised after the device is modelled. Implications of modern medical imaging, mechanical and computational methods influence biomedical device product research and development. For example, in the later part of the book, we will examine how these medical image-based computational models can be utilised in simulation packages, applications and testing. We will also integrate the medical imaging methods discussed in Chapter 4. Chapter 5 examines the aneurysmal treatment based on stenting techniques, and further elaboration on endovascular stent grafts are provided in Chapter 6. Chapter 7 and Chapter 8 give an overview of the nasal drug delivery and biomedical MEMS micropump technology, respectively. Finally, the description of the generic biomedical device production organisation based on scientific management, industrial and strategic organisation concepts, as well as product development system implementation are concluded in Chapter 9.

Acknowledgments

This book emerged from the support rendered by our fellow colleagues and friends. The many conversations and interactions with them has guided us through our writing and organisation of its contents. Through the times that we spent in preparing this work, we have also built a good working relationship and friendship with our peers who have contributed directly or indirectly to our book. We would like to thank the research students and colleagues at the School of Aerospace, Mechanical and Manufacturing Engineering, RMIT University, for their kind effort in various ways towards the compilation of this work. In particular, special thanks are extended to Dr. Sherman Cheung for his useful comments on the aspects of computational haemodynamics and experimental validation. We appreciate Dr. Kiao Inthavong for his contribution on experimenting the nasal drug delivery techniques, as well as producing the related graphics and results that appear in Chapter 7 of this book. There is also appreciation of the help rendered by Mr. Thanapong Chaichana from the Department of Imaging and Applied Physics, Curtin University, for his assistance in preparing the diagrams, charts and graphical information relating to Chapters 4, 5 and 6. We are grateful to Prof. Joon Hock Yeo from the School of Mechanical and Aerospace Engineering, Nanyang Technological University, for the diagrams of the prosthetic heart valves and illustration of their design principles that appear in Chapter 3. Next, appreciation is extended to Dr. Said Al-Sarawi from the School of Electrical and Electronic Engineering, University of Adelaide, for his kind advice on the Bio-MEMS research materials presented in Chapter 8. Last but not least, our gratitude to Ms. Veronica Ling Hui Low from World Scientific for her time spent in discussing the style of presentation and format of the contents in this book.

Contents

Foreword v

Preface ix

Acknowledgments xi

1 Introduction 1
- 1.1 Overview of Research and Development Processes 1
- 1.2 Questions 6

2 Overview of Biomedical Technologies 7
- 2.1 Classification of Biomedical Devices 7
- 2.2 Description of Biomedical Devices 7
  - 2.2.1 Aneurysmal Stents 9
  - 2.2.2 Endovascular Stents 10
  - 2.2.3 Biomedical MEMS Micropump 11
  - 2.2.4 Drug Delivery Devices 13
- 2.3 Summary 15
- 2.4 Questions 15

3 Conceptualisation and Medical Image-Based Modelling 17
- 3.1 CAD Modelling and Design Realisation 17
  - 3.1.1 Prosthetic Heart Valve 17
  - 3.1.2 Endovascular Stent Grafts 18
  - 3.1.3 Biomedical MEMS Micropump 19
- 3.2 Medical Imaging and Reconstruction 23
  - 3.2.1 Computed Tomography 24
  - 3.2.2 Virtual Intravascular Endoscopy 25
  - 3.2.3 CT Reconstruction of the Nasal Cavity, Pharynx and Larynx 26
  - 3.2.4 Magnetic Resonance Imaging 28
- 3.3 Mechanical Prototyping 29
  - 3.3.1 Rapid Prototyping by Stereolithography 30
  - 3.3.2 Technical Limitations 32
- 3.4 Summary 34

3.5 Questions 35

4 Medical Imaging and Visualisation **37**
4.1 Computed Tomography 37
4.2 Virtual Intravascular Endoscopy 38
4.2.1 Generation and Presentation of VIE 41
4.2.2 Generation of VIE Images 42
4.2.3 Threshold Range Along the Abdominal Aorta 42
4.2.4 Optimal Threshold Selection 43
4.2.5 Generation of VIE Images with Aortic Stent and Artery Lumen Together 44
4.2.6 Aortic Stent Wire Thickness on VIE Images 47
4.2.7 Image Display and Interpretation 47
4.3 Optimal CT Scanning Protocols for VIE Visualisation 51
4.4 Summary 54
4.5 Questions 54

5 Treatment of Aneurysms **55**
5.1 Introduction 55
5.2 Open Surgery 56
5.3 Minimally Invasive Techniques 57
5.4 Medical Image Visualisation 58
5.5 Technical Limitations 60
5.6 Medical Imaging and Geometrical Reconstruction 60
5.7 Conformance with Preliminary Concept 65
5.8 Summary 66
5.9 Questions 67

6 Endovascular Stent Grafts **69**
6.1 Review of Device 69
6.1.1 What Is a Stent Graft? 70
6.1.2 Why Endovascular Repair? 71
6.2 Technical Developments 71
6.2.1 Suprarenal Stent Grafts 72
6.2.2 Fenestrated Stent Grafts 73
6.3 Technical Success 76
6.4 Long-term Outcomes 76
6.5 Computational Modelling 78
6.5.1 CFD of Suprarenal Stent Grafts 78
6.5.1.1 Configuration of Stent Wires Crossing the Renal Artery Ostium 79
6.5.1.2 Segmentation of CT Volume Data 79
6.5.1.3 Generation of Aorta Mesh Models 80

6.5.1.4 Simulation of Suprarenal Stent Wires Crossing the Renal Artery Ostium . . . 81
6.5.1.5 Computational Two-Way Fluid Solid Dynamics . 83
6.5.1.6 CFD Analysis . . . 84
6.5.2 CFD of Fenestrated Stent Grafts . . . 84
6.5.2.1 Simulation of Fenestrated Renal Stents . . . 86
6.5.2.2 Numerical Verification . . . 86
6.5.2.3 Computational Two-Way Fluid Solid Dynamics and Analysis . . . 86
6.6 Summary . . . 87
6.7 Questions . . . 88

7 Nasal Drug Delivery 89
7.1 Review of Device . . . 89
7.2 Computational Modelling . . . 93
7.2.1 Geometrical Meshing . . . 93
7.2.2 Physiological Boundary Conditions . . . 95
7.2.3 Simulating Flow in the Nasal Cavity . . . 97
7.3 Assessment of Modelling and Optimisation . . . 101
7.3.1 Insertion Angle . . . 101
7.3.2 Full Spray Cone Angle . . . 103
7.3.3 Implications for Nasal Drug Delivery . . . 107
7.4 Summary . . . 107
7.5 Questions . . . 110

8 Biomedical MEMS Micropump 111
8.1 Review of Device . . . 111
8.2 Biomedical Applications of MEMS Micropumps . . . 112
8.2.1 Potential Drug Delivery Applications . . . 112
8.2.2 Other Potential Applications . . . 113
8.3 Numerical Modelling of Micropumps . . . 114
8.4 Operating Principle of an Example Micropump . . . 115
8.5 Theoretical Analysis . . . 115
8.5.1 Actuation Force . . . 117
8.5.2 Microfluidic Pressure Variation . . . 117
8.6 Model Development of an Example Micropump . . . 118
8.7 Modelling and Simulation . . . 119
8.8 FEA-based CFD Simulations . . . 121
8.8.1 Inlet and Outlet Flow Characteristics . . . 123
8.8.2 Flow Rate Estimation . . . 124
8.9 Summary . . . 126
8.10 Questions . . . 126

9 Engineering and Production Management for Biomedical Devices 127
9.1 Overview of Product Development Strategy 127
9.1.1 Product Conceptualisation 129
9.1.2 Market Survey and Strategic Alliances 129
9.1.3 Design, Prototyping and Product Development 130
9.1.4 Testing and Commissioning 132
9.1.5 Technology Protection 132
9.1.6 Marketing and Sales 133
9.2 Product Development Management System 134
9.2.1 Systematic Management 134
9.2.2 Standard Operation Procedures 134
9.2.3 Material Resource Planning 135
9.3 Organisation Charts 135
9.4 Project Management 139
9.4.1 Fundamentals of Project Management 140
9.4.2 Project Life Cycle 142
9.4.3 Project Management Team 143
9.5 Summary 146
9.6 Questions 147

Bibliography 149

Answers to Chapter Questions 169

Index 175

1

Introduction

CONTENTS

1.1 Overview of Research and Development Processes 1
1.2 Questions 6

1.1 Overview of Research and Development Processes

Medical research is the iterative search for the understanding of the human body based on a scientific process. Such pure research seeks for the truth without a pre-defined utility at the start, whereas applied research that pertains to the development of biomedical devices seeks to solve medical related problems with the motivation of addressing the patient (or consumer) needs. Devices arising from applied research can be used to support pure research. For instance, the developed biomedical devices such as medical image scanners can be used to examine the human body physiologically and conduct medical research. This book focuses on the science of research and development of biomedical devices. In recent decades, the advancement of medical imaging, computer technology and manufacturing processes have added value the conceptualisation and development of new biomedical applications, and significantly improved health and living standards as well as contributing to medical science and technology.

An example of a biomedical device may be a surgical assisting application, or a mechanism of delivering drug into the human body. The purpose of using such a device is to provide specific medical assistance and long-term safety while minimising the device-related complications. The presentation of a systematic framework addresses the development of a biomedical device to meet health standards or safety limits by showing the integration of the required processes and instruments used in its design, evaluation, prototyping and delivery. In this book, we introduce the outline of devising and testing some selected biomedical devices as well as to highlight the nature of their applications and impact to the medical industry.

There has been a rapid introduction of computational modelling tools, medical imaging, optical-based flow measurements, sensor technology and manufacturing processes into the development of a biomedical device. This has resulted in the implementation of the device from the start of its conceptualisation to prototyping and testing to become an engineering challenge and requires detailed management and analysis at every stage of its development.

The first stage of biomedical device development is design realisation and innovation. Design rules based on certain specifications defined along the process of realising the product can be achieved with reference to a set of product operation and safety guidelines. Two types of design innovation exists: *autonomous* – design realisation independent from its other existing innovations, whereby a new design rule to increase the effectiveness of the device or its level of safety can be developed without redesign of the entire product; and *systemic* – design realisation in conjunction with its related complementary innovations, whereby new design updates requires the need to reformulate the old design rules and redefine the design of the product. Coordinating an *autonomous* or *systemic* design product innovation is particularly difficult when dealing with many design options. Therefore, computational modelling and design based on computer-aided design packages can help to lower product development cost significantly. Alternatively or as a complement platform, rapid prototyping based on stereolithography[1] can also assist biomedical engineer in visualising their products or to be utilised during experimental testing.

The next stage is the visualisation of the product performance, which can be computational or mechanical simulation, and in some cases, based on clinical imaging. Computational tools such as computer-aided design (CAD) software, numerical solvers for computational fluid dynamics (CFD) and flow visualisation packages are mainly used in simulation, visualisation and verification stages of the development. Medical imaging such as magnetic resonance imaging (MRI) and computed tomography (CT) can produce images of anatomical or functional features *in vivo*. These images can be processed and reconstructed to provide useful information for visualisation using computer modelling tools. Medical imaging is well known for its ability to present the anatomical scenario within the human body so that identification of locations for device implants, drug delivery or rectification of anatomical defects may be examined.

As mentioned, the evaluation of the device performance may be experimental, computational and clinical. Figure 1.1 depicts the different computational and mechanical instruments in the development cycle of a biomedical device that play specific roles. Experimental simulations using particle image velocimetry (PIV) and laser Doppler measurements have been able to derive fluid mechanical properties of flow through mechanical devices, such as the prosthetic heart valves and stents, before actual implantation into the human body. Such experimental information may often be complemented by CFD simulation for cross-checking. In some cases, it may be possible to retrieve *in-vivo* functional measurements for clinical verification. For example, phase contrast MRI may be used to present flow maps of the cardiac circulation in the heart and compare against those that are simulated.

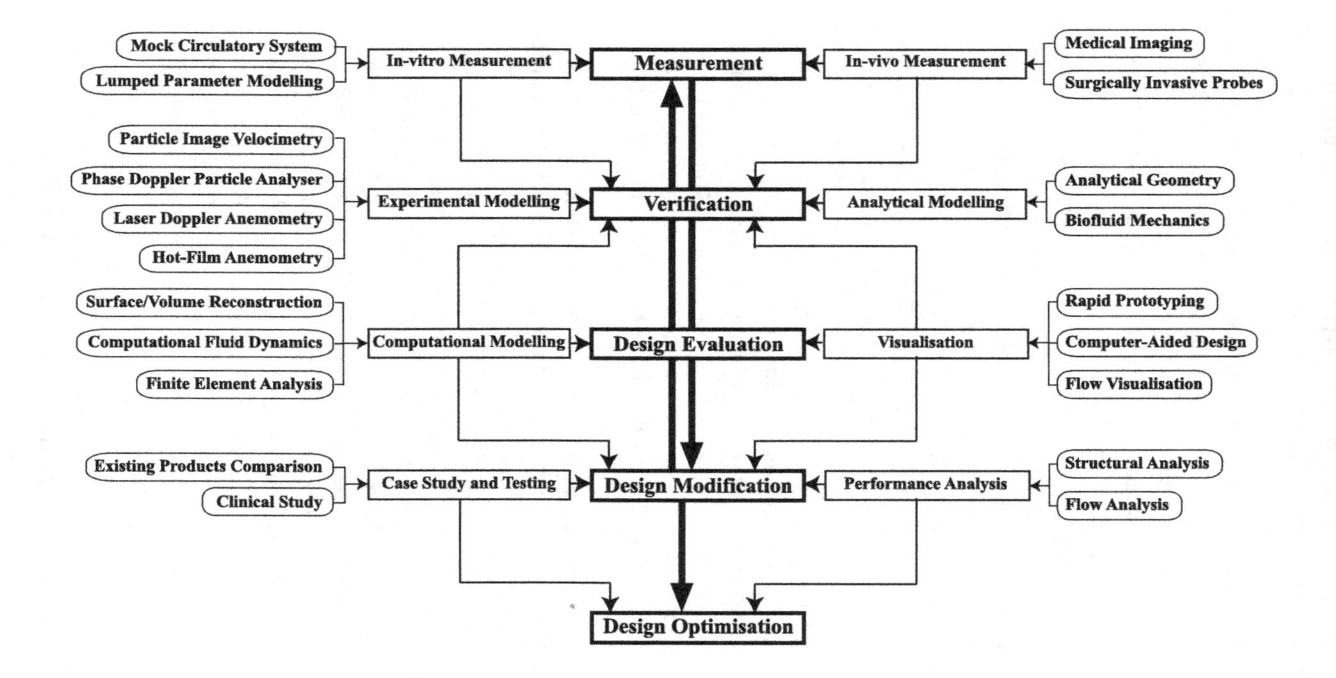

FIGURE 1.1
Stages for developing a biomedical device. The success of a biomedical device involves various stages in its design evaluation and testing from the structural and flow perspective. The anatomical and functional characteristics of the pathology prior and post surgical implantation of a biomechanical structure can be examined for its effectiveness. The simulation of drug delivery process based on various mechanical parameters can assist in improving the device design. These testing and remodelling processes involve the system integration of various technical developmental procedures, which can be computational or experimental.

We examine the design and measurement, prototyping, testing, evaluation and proposal for delivery of a biomedical device from the system engineering perspective. Some of these processes may be performed on the mechanical or computation platform. In the generic sense, the terminology measurement refers to the physical scanning of parameters in the system evaluation of the device. The term computational means numerical calculations with the aid of a computerised system while mechanical relates to the physical extraction of data or modelling of a hard-copy prototype in terms of biomedical product development.

Ideally, *in-vivo* medical imaging and *in-vitro* measurement devices may be used for modelling of the actual device performance before and after its implantation depending on the availability and permission for experimentation of the test subject. However, this may sometimes be limited by the ability to measure the device in operation due to technological constraint. For example, nuclear signals from metallic device show up as void in MRIs.[2] *In-vivo* measurements give a perspective of the realistic impact that the biomedical device can have on the anatomical functions whereas *in-vitro* measurements, often used in experimental verification of the device, are based on a mechanically confined environment where the parameters of the anatomical events may be simulated or the introduction of the measurement sensors may affect the performance of the device *in situ*. Computational modelling may sometimes supplement the measured data as a form of cross-check. As such, measurements together with the computational modelling can provide a strong examination of the biomedical operations with the relevant anatomical functional behaviour.

Visualisation is important after the measurement process. Collected data needs to be processed and presented in useful formats to the user. For example, the geometrical determination of cardiovascular or respiratory structures by medical imaging or scanning in two dimensions (2D) needs to be reconstructed in a computational platform for three-dimensional (3D) visualisation. This can assist in the evaluation of design and also facilitate the simulation of device performance computationally once the anatomical environment is known more clearly. With the rapid development of medical imaging techniques and visualisation tools, 3D reconstructions offer additional information when compared to planar axial images. For example, in pre- and post-endovascular stent grafting, it has become a routine clinical tool in both surgical planning and post-operative follow up.[3,4] This is especially demonstrated by virtual intravascular endoscopy (VIE) visualisation that can allow an accurate assessment of the treatment outcomes of endovascular aortic stents in terms of the intraluminal appearances and stent position or protrusion.

Experimental flow mapping may be broadly classified with the following main characteristics and nature: optical, magnetic resonance or ultrasonic-based velocimetry systems.[5] PIV is an example of an experimental optical velocimetry system used to verify design performance. The design of a device or structure can be investigated experimentally using a physical model which is interrogated using techniques such as PIV performance.[6,7]

Velocity-encoded medical imaging such as phase contrast MRI can be used to generate 2D flow maps. Of recent advancement is the phase contrast MRI that is able to map flow fields of up to 3D *in vivo*.[8–10] Two-dimensional ultrasonic flow field mapping has been able to enable useful results in cardiac diagnosis,[11] and can potentially be developed into an experimental flow imaging system for biomedical device evaluation.

Finite element analysis (FEA) is another category of computational simulation that is able to evaluate structural performance of the device. Considering the ability to model more advanced geometries, FEA has become a popular computational numerical method, oppose to theoretical or analytical modelling. Various researchers have utilised advanced FEA techniques for modelling and analysis of implantable devices for applications such as drug delivery systems,[12,13] artificial organs[14–17] and endovascular stents.[4]

CFD is a numerical method that can be used to provide detailed flow information in the human that cannot easily be provided by experiment.[18–22] Furthermore, CFD and FEA techniques complement results to experiments. Once validated, the model can be used to investigate the effects of changing parameters or geometry with greater certainty, and at substantially less cost than building a new experimental prototype. In addition, the simulation of design performance can be achieved to evaluate the safety of the device implant or the effect of introducing drugs into a system without endangering human lives. It may serve to provide expert opinion to surgeons in the event of strategising the device implant through better understanding of its operating mechanism.

During design evaluation, existing biomedical device products are compared and calibrated against the proposed design in terms of performance. In this analysis, statistical studies based on device testing on a population of human subjects are an important criterion in commissioning the product. The flow performance of the device is also another consideration as improved operation of the product will enable better marketability. Products often need to be designed and re-designed again based on the vigorous testing and evaluation procedures.

In summary, in order to sufficiently evaluate the design of a biomedical device such that we are to achieve efficient, effective and safe operation during surgical implant or clinical usage, the implementation of measurement, verification, design evaluation and modification should be the standard operating procedures during its product development. The use of both physical "measurement" and simulated "computational" information are equally important, and should be used to complement the verification of the testing and modelling techniques sufficiently. These may be supported with various techniques in the system of design and testing. For instance, the use of PIV has been widely used to verify CFD results in device testing and also in drug delivery performance.[6,7] There have also been cases of medical imaging measurement methods such as phase contrast MRI that is used to verify CFD results in various physiological studies.[23] Then, for production of the medical technologies, the engineering and manufacturing of the biomedical products require a set of good scientific management principles in an organisation that undertakes its mass production and commercialisation. Our book covers all aspects of the aforementioned methods used for research and development as well as the production of biomedical products.

1.2 Questions

1) What are the stages leading to the research and development of biomedical devices?
2) What is/are the difference(s) between pure research and applied research?
3) What are the two categories of design innovations?
4) Name two methods that are used during the verification stage of biomedical device development and list example technologies for each method.
5) Name the broad categories of experimental flow velocimetries.

2

Overview of Biomedical Technologies

CONTENTS

2.1 Classification of Biomedical Devices 7
2.2 Description of Biomedical Devices 7
    2.2.1 Aneurysmal Stents 9
    2.2.2 Endovascular Stents 10
    2.2.3 Biomedical MEMS Micropump 11
    2.2.4 Drug Delivery Devices 13
2.3 Summary 15
2.4 Questions 15

2.1 Classification of Biomedical Devices

As depicted in Fig. 2.1, a biomedical device can be classified into five categories that serve both diagnostic and therapeutic purposes. Various types of drug delivery mechanisms such as nasal, gastrointestinal and intravascular drug delivery systems are available for a variety of applications. Surgical restoration device such as the stent is mainly used to restore normal blood circulation through implantation of biomedical device into the affected tissues or organs. The stent can serve to prevent cardiovascular-related complications such as re-stenosis or occlusion in arteries.

Medical-assisting devices are applied to correct pathological conditions and applications include cochlear implants, corrective lenses and heart–lung machines. Diagnostic devices refer to both procedures such as blood pressure and glucose monitoring to provide measurement and diagnosis of clinical situations. Mechanical part replacement indicates the use of artificial components to substitute the diseased anatomical structures or organs which cannot be treated medically.

2.2 Description of Biomedical Devices

In this section, we present the properties or characteristics of the device, which includes the mechanism of operation and evaluation techniques. Each of the following

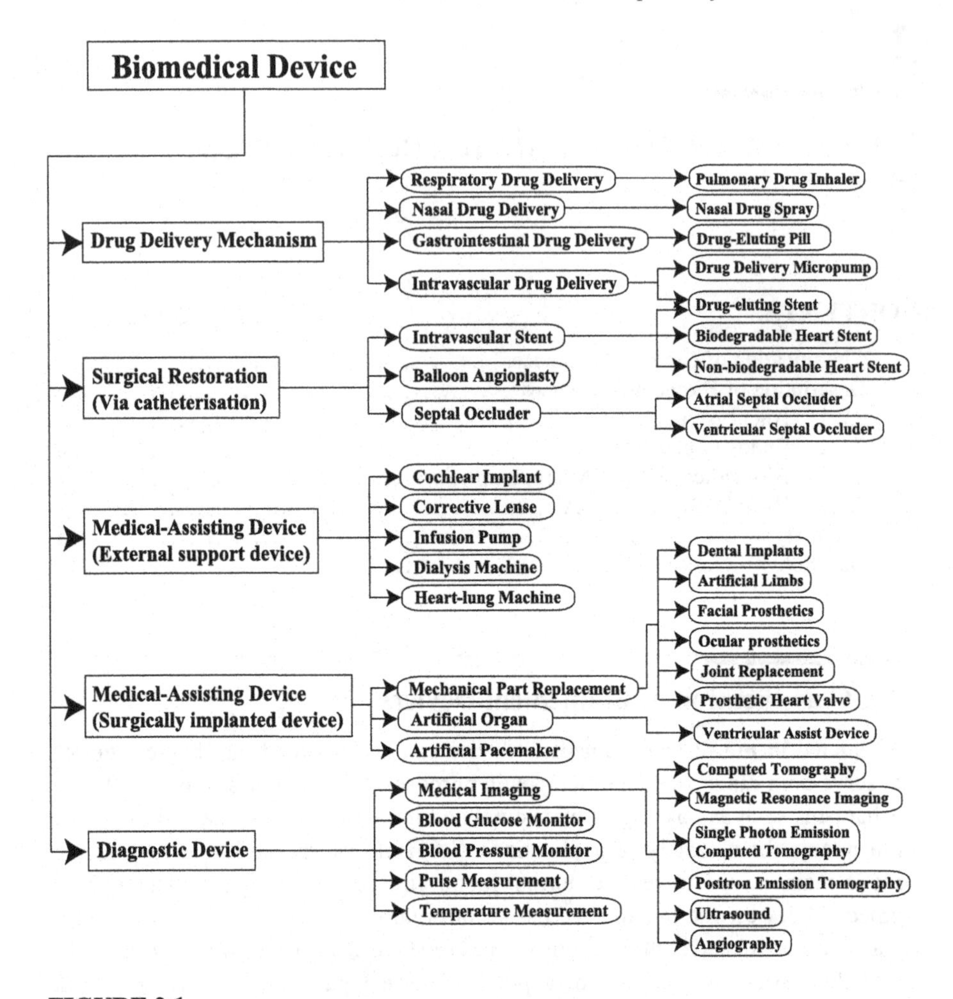

FIGURE 2.1
Characterisation of medical technologies. A biomedical device can be classified into five categories which serve both diagnostic and therapeutic purposes: (1) drug delivery mechanisms aim to optimise the transport of medicine into the body; (2) surgical restoration devices are mainly used to restore normal a normal body function through implantation of biomedical device into the affected tissues or organs; (3) medical-assisting devices are applied to correct pathological conditions; (4) diagnostic devices refer to both medical imaging and non-medical imaging procedures to provide measurement and diagnosis of clinical situations; and (5) mechanical part replacement indicates the use of artificial components to substitute the diseased anatomical structures or organs which cannot be treated medically and need to be amputated.

subsection highlights of the nature of a biomedical flow device and the considerations related to its testing and evaluation.

2.2.1 Aneurysmal Stents

Aneurysm is defined as the focal dilatation in the arterial wall and is commonly caused by atherosclerotic disease of blood vessels. With aging, aneurysms increase in size, which results in the rupture and initiation of bleeding within the body regions, such as the brain. This is responsible for high mortality such as haemorrhagic stroke. The affected arteries are generally treated with stents through angioplasty procedures. Medical imaging techniques such as CT and conventional angiography play an important role in the detection of aneurysms and preoperative planning. To achieve effective treatment, the type of stents deployed has to be of sufficient porosity to minimise aneurysmal rupture,[24] and yet has to prevent platelet aggregation. A computational model can be simulated to show stenting inside the diseased artery at the entrance of the aneurysm, and demonstrate that the interference with blood flow is different from the situation before stent treatment. Figure 2.2 shows the various stent porosities that can be configured for an aneurysm.

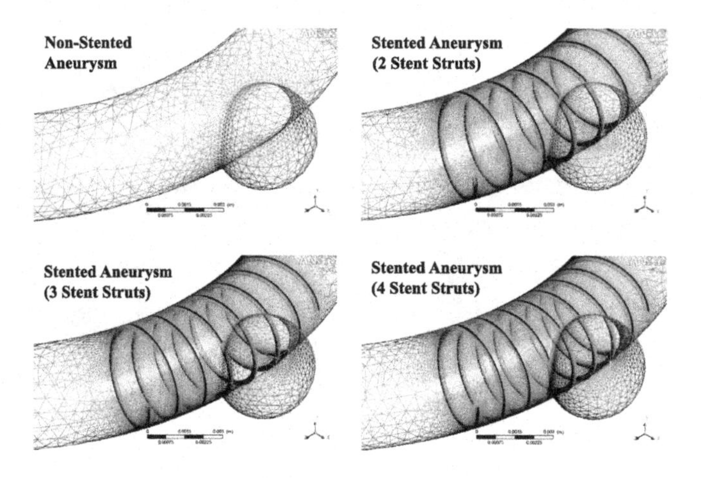

FIGURE 2.2
Configuration of stent porosities. The reason to insert stents inside the aneurysm is to embolise it or exclude it from systemic blood circulation, so that the aneurysm will shrink gradually (due to reduced pressure) and eventually become smaller. It is true that this may induce stagnation and possible thrombosis formation after stenting; however, as long as the aneurysm does not grow, the chance of being ruptured is low, so the goal of preventing it from rupture is achieved.

However, no haemodynamic information is available with these image visualisation tools. The CFD technique is increasingly used to quantify blood flow through pre- and post-stented aneurysms defined from medical imaging techniques. Based on numerical simulation, intra-aneurysmal flow before and after stenting can be studied, and vorticity mapping can be performed to enable analysis of the large-scale vortex in the aneurysm sac. Superior haemodynamics by a reduced inflow and intra-aneurysmal velocities can be assessed using a stereo PIV platform in order to provide a concise experimental insight into the aneurysmal haemodynamics.[25] This assists clinicians in developing stents that match the patient-specific vessels and improve the treatment outcomes.

2.2.2 Endovascular Stents

Endovascular stent grafting was first introduced into the clinical practice to treat abdominal aortic aneurysms (AAAs) as a less invasive technique in the early 1990s.[26] Since then, the research and development of stent grafting is propagated by medical interests and its clinical utilities.[27–29] The suprarenal and fenestrated endografts have been engineered to treat patients presented with short and angulated aneurysms necks, which cannot be resolved with conventional infrarenal stent grafts.[30–32]

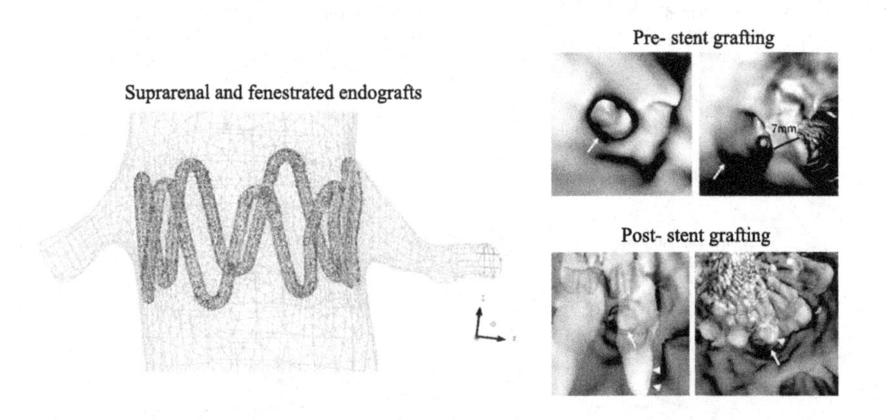

FIGURE 2.3
Virtual endoscopy (VE) views of pre- and post-stent graft implantation. VE views are generated to visualise the ostia of bilateral renal arteries (arrows). VE provides superior information about the intraluminal appearance of coronary wall when compared to conventional external views regarding the surface and configuration of the renal artery ostium. Also, the aneurysm neck can be accurately measured intraluminally for planning of the stent graft implantation. Thereafter, post-stent grafting results can be visualised by the same technique.

While stents and stent grafts are increasingly used in clinical practice as an effective alternative to open surgery, the long-term outcomes are yet to be determined. One of the main concerns of implanting such devices into the blood vessels is thromboembolism or re-stenosis. Another potential risk is the interference of regional blood flow by the implanted devices. Research in this area is being performed, and it has been reported that the placement of stents alters the haemodynamics and this coupled with wall movement may lead to the dispersion of late multiple emboli.[33] The complex structures that are introduced into the blood flow (renal blood flow in the fenestrated stent grafts) may enhance the biochemical thrombosis cascade[34] as well as directly affect the local haemodynamics. Further studies are still needed to verify the safety of these biomedical devices using a combination of 3D visualisation and CFD. An in-depth investigation of the effectiveness of these biomedical devices is valuable to improve the design of the devices, treatment outcomes and reduce the potential risk of developing procedure-related complications, such as ischaemic changes or re-stenosis or occlusion. Figure 2.3 is a 3D virtual endoscopy (VE) image showing the intraluminal view of renal artery ostium in a patient before and after endovascular stent grafting.

2.2.3 Biomedical MEMS Micropump

The number of patients suffering from life-threatening global health problems and chronic diseases such as cardiovascular deceases (hypertension, heart attack, stroke), septicaemia, cancer, diabetes, melancholia and malignant lymphoma have become greater than ever before.[35] Medical and biological experts view that early recognition and treatment are the key to eliminate and/or control such risky disorders.[36,37]

Conventional drug delivery methods such as oral tablets or injections have various limitations. Among them, the problems of variable absorption profiles and the need for frequent dosing are yet to be addressed.[35] Furthermore, conventional controlled release formulations are designed to deliver drugs at a predetermined, preferably constant, rate. Some clinical situations, however, necessitate either an external control of the drug delivery rate, or a volume of drug that is beyond the capabilities of existing controlled release formulations.[35] Therefore, conventional methods are not effective in delivering a drug within their therapeutic range.[38,39] Moreover, most of the currently existing biomedical implants extract energy from an implanted battery, which inherently limits its operating lifetime due to their fixed energy density and the strict volume and mass constraints imposed by implantable devices. Hence, a new alternative transcutaneous energy transmission system is also highly regarded for microdevices that are targeted at *in-vivo* applications. It is evident that effective techniques are required for accurate and effective early diagnostics and/or targeted treatments. Such techniques should incorporate automatic dosing capabilities, so that the patients are prevented from sudden death or irregular/incorrect intake of medicine/drug. Therefore, the implementation of targeted micro-drug delivery methods is recognized as one of the critical solutions in the life-saving health advances.

A bio-micro electromechanical systems (Bio-MEMS)-based typical micropump is a fundamental part of an implantable drug delivery system, which provides the actuation source to effectively dispense drugs or therapeutic agents to a targeted area in the body with precision, accuracy and reliability. It is dependent on the operation of a microdiaphragm that moves up and down to push biofluids through a channel (Fig. 2.4).

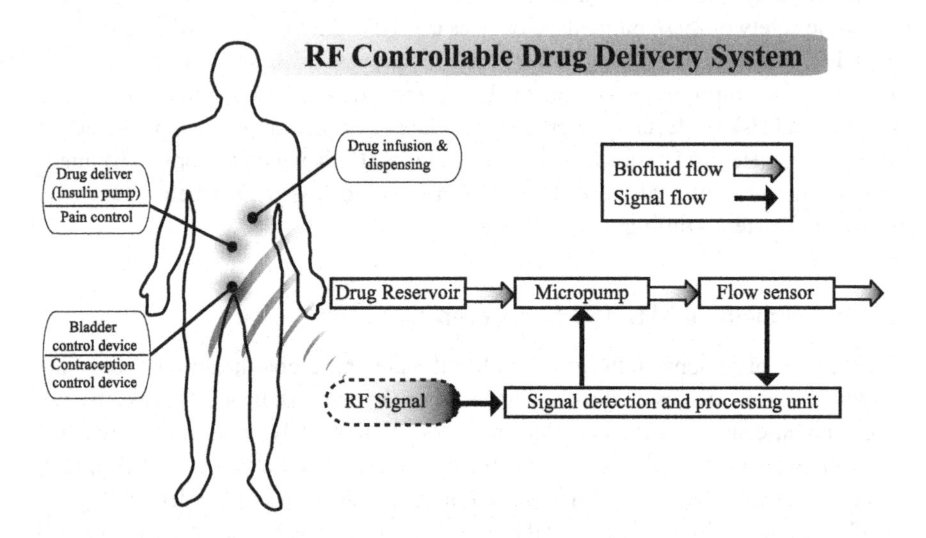

FIGURE 2.4
Biomedical MEMS Micropump. A radiofrequency (RF)-controllable microvalve or a micropump is the next generation of intelligent implantable devices in the human body. This concept exploits the principle of a surface acoustic wave (SAW) device on a piezoelectric substrate, and possesses high potential in complimenting or replacing biological functions from the micro-level perspective. Such a biocompatible micropump could be implanted for applications such as drug infusion and dispensing/delivery and non-hormonal contraception. Such a device comprises a drug reservoir, micropump, micro-flow sensors, micro-channels with valves and RFID-type secure mechanisms that wirelessly interrogate biofluid flow control.

Micropumps are also an essential component in fluid transport systems such as micrototal analysis systems, point-of-care testing systems and lab-on-a-chip (LoC). These devices are used as a part of an integrated LoC consisting of microreservoirs, microchannels, microfilters and detectors for the precise movement of chemical and biological fluids on a micro-scale. The point-of-care testing system is a micrototal analysis system to conduct diagnostic testing on site close to patients to provide better health care and quality of life. In such diagnostic systems, MEMS micropumps are integrated with biosensors on a single chip to realise smart devices. Such micropumps typically relies on various actuation mechanisms such as electrostatic, piezoelectric, electromagnetic, thermo-pneumatic, shape memory alloy (SMA) and bimetallic.

Micropumps or microvalves that are built on the concept of wireless transcutaneous radiofrequency (RF) communication in biomedical applications can generate a *remotely controlled* device for control of fluid transport.[12] Commercial software such as finite element modelling (FEM)-based CFD tools (ANSYS, CFX and Coventor) and mechanical event simulation (ALGOR) can be utilised to simulate micropump motion and flow rectification in its design optimisation.

2.2.4 Drug Delivery Devices

The human respiratory system generally comprises the lungs, conducting airways, pulmonary vasculature, respiratory muscles and surrounding tissues and structures. Each of these plays an important role in influencing respiratory responses. The nasal airway is dominated by nasal turbinates that are lined with highly vascularised mucosa that contain openings to the paranasal sinuses. Due to these characteristics, it is hypothesised that drug delivery to combat health problems such as lung diseases, cancers, diabetes, sinus infections, etc. may be viable if the drug formulation can be deposited in the turbinate region.[40] However, current nasal delivery devices have major disadvantages, one of which is the large proportion of the drug particles deposited in the anterior regions of the nasal vestibule — attributed to the sprayed particles existing in a high inertial regime.[1,7,41] Figure 2.5 shows how a nasal drug delivery spray looks like.

Experimental evaluation of the nasal spray device for delivery of drugs can be performed with human volunteers through the use of gamma scintigraphy[42] or cast models of the nasal cavity itself.[43] In the case of human volunteers, the option of having the user actuate the spray themselves can introduce inter-individual variances. For instance, an elderly person or young child may not be as physically strong as an adult, which leads to different actuation strengths and speeds, which eventually lead to variations in spray atomisation. This is highlighted in Refs. 44 and 45.

Other research and development in this field have been aimed at understanding some of the parameters (formulation viscosity[46] and spray actuation by the user[47,48]) that affect the ability of the spray device in order to deliver high drug efficacy. Evaluation of the spray device in these studies is measured by external spray characteristics such as the droplet size, plume geometry and spray pattern. These tests were perhaps chosen based on the U.S. Food and Drug Administration (FDA) draft guidance, which propose that the bioavailability and bioequivalence of nasally administered, locally acting drug solutions may be determined solely using *in-vitro* methodology. However, Suman *et al.*[42] showed that the spray pattern test did not provide any conclusive evidence towards actual deposition patterns. In light of this, the evaluation of nasal spray devices cannot be solely relied on one method, but rather a synergistic approach is most effective by combining *in-vitro* and *in-vivo* experimental and computational methods.

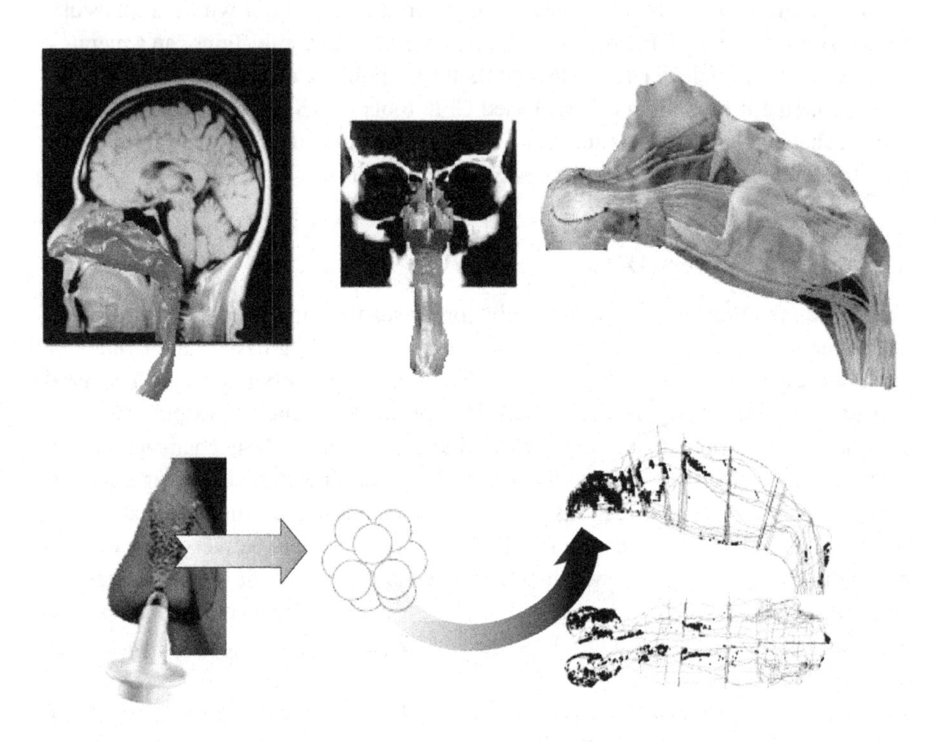

FIGURE 2.5
Nasal drug delivery spray device and therapy testing. Inhalation of drug particles deposited directly to the lung periphery results in rapid absorption across bronchopulmonary mucosal membranes and the reduction of adverse reactions in the therapy of asthma and other respiratory disorders. As such, it is desirable to design nasal sprays that deliver the particles such that they are not deposited in the upper airways before reaching the lung periphery. This is because excessive deposition of drug particles in the upper airways will cause diminished therapeutic effects in the lung or local side effects in the upper conducting airways.

2.3 Summary

In this chapter, we selected some of the devices that pertain to the drug delivery mechanism, surgical restoration device, and mechanical part replacement categories to demonstrate the development philosophy and key conceptual processes that relate to its generation. In particular, we have described these computational and mechanical techniques applied onto the various stages of developing three categories of flow-related biomedical devices. The (1) prosthetic cardiac devices, (2) cardiac restoration devices and (3) drug delivery mechanism are investigated as an illustration of the development and evaluation processes involved.

2.4 Questions

1) Name the categories of biomedical devices.
2) What is an aneurysm? Name one type of treatment for this disorder.
3) Name three applications of the biomedical MEMS micropump.
4) Name five different actuation mechanisms of the biomedical MEMS micropump.
5) Name one major disadvantage of the existing nasal delivery device.
6) Explain why it is desirable avoid drug deposits in the upper airways before reaching the lung periphery.

3

Conceptualisation and Medical Image-Based Modelling

CONTENTS

3.1 CAD Modelling and Design Realisation 17
3.1.1 Prosthetic Heart Valve 17
3.1.2 Endovascular Stent Grafts 18
3.1.3 Biomedical MEMS Micropump 19
3.2 Medical Imaging and Reconstruction 23
3.2.1 Computed Tomography 24
3.2.2 Virtual Intravascular Endoscopy 25
3.2.3 CT Reconstruction of the Nasal Cavity, Pharynx and Larynx 26
3.2.4 Magnetic Resonance Imaging 26
3.3 Mechanical Prototyping 29
3.3.1 Rapid Prototyping by Stereolithography 30
3.3.2 Technical Limitations 32
3.4 Summary 34
3.5 Questions 35

3.1 CAD Modelling and Design Realisation

This section is more specific on the modelling of the biomedical device and its related physiological environment. We illustrate the prosthetic heart valve, endovascular stent and biomedical MEMS micropump as case examples for device conceptualisation and modelling. Later on, we will also explore the medical imaging techniques used to examine the anatomy that presents the environment for the device.

3.1.1 Prosthetic Heart Valve

Mechanical heart valves are relatively durable but are strongly associated with thromboembolisms, which often result in ischaemic attacks and strokes. Bio-prosthetic heart valves come with a low incidence of thromboembolism; however, there is stiffening of the tissue due to the build up of calcium. Flow visualisation of these two types of heart valves can provide useful information on the mechanical and operational characteristics of these devices. Computational and experimental flow visualisation of heart valves can be performed; however, there is a need to prepare the structural model on the drawing board before prototyping. The experimental setup using PIV or LDA will enable us to construct useful flow maps of the space surrounding the heart valve, and engineers typically prototype a scale-up physical model to be used in the mechanical test environment. Advances in CFD have assisted greatly in evaluating the hydrodynamic performance and structure function correlations in heart valves[49,50] by modifying various parameters of the device computationally with minimal material resources. The measured data can be used to verify numerical simulations of the flow for a typical mechanical or bio-prosthetic heart valve. We briefly identify some of the technologies used in heart valve modelling.

The flow characteristics of the blood distal to the aortic heart valve are measured from clinical experimentations and medical images. The mechanism of flow is separated into rotational and translational components and the magnitude of each is determined. The same flow mechanism is duplicated upstream of the valve in a simulation platform, and the flow that results from this setup will be compared with those of the experimental and existing clinical data.

Mechanical heart valves are typically easier to model unlike non-deformable tissue valves. However, their designs can be particularly challenging. For instance, a bad design may simulate thrombosis resulting in ineffective operation of the tilt discs, and jeopardise the opening and closing of the valve. Therefore, there are certain design principles to follow when developing an effective heart valve. An example of a bileaflet tilt disc mechanical and tissue heart valve is shown in Fig. 3.1.

Computational modelling of tissue heart valves can be produced using CAD software packages that range from 2D vector-based design platforms to 3D solid or surface modelling systems (Fig. 3.2). Special materials are used to construct the tissue heart valves. Based on the different structural nature of the leaflets, the design of the device requires intensive testing using computational as well as experimental techniques. The models may assist engineers in the design and evaluation of the mechanical structures that pertain to the heart valve.

3.1.2 Endovascular Stent Grafts

CT imaging with the aid of VIE can generate the 3D reconstruction of aortic branches and anchoring stents to evaluate how the stent grafts situate within the aorta after placement of stent grafts.[3,51] Figures 3.3 and 3.4 illustrate the stent placement in the aorta in a simulated environment.

The use of computational tools such as a 3D meshing software allows the modelling of an anatomy based on the surface reconstruction using connected points on coronal-sectioned scans. A volume mesh or grid can be generated and improved by cell adaptation techniques to refine the large volume cells that pertain to high velocity

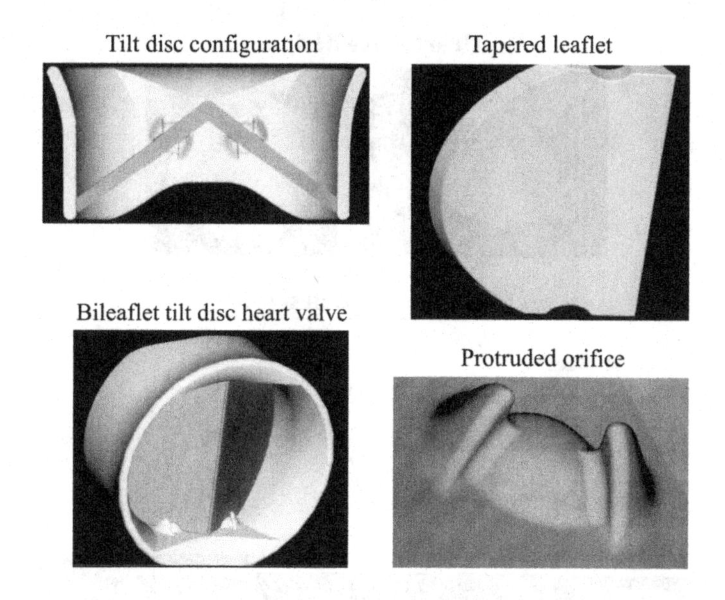

FIGURE 3.1
Bileaflet tilt disc heart valve mechanical design. Computer-aided design (CAD) was used to develop a heart valve that comes with a tapered leaflet, a bell-mouth orifice for minimum pressure drop across the valve and a protruded ellipsoidal hinge. The purpose of the tapered leaflet is for minimal impact on the orifice, and the protruded orifice is for the positive washing of hinges to enhance the performance of the heart valve as shown by their design configurations. (Courtesy of Prof. Joon Hock Yeo, Nanyang Technological University, Singapore.)

gradients. Near-wall grid refinements are also performed to improve the convergence of solutions for the conservation equations of mass and momentum.

In computational-based modelling, various features and capabilities are available to appropriately mesh a design. The accuracy of computational analysis in the computational platform depends on factors such as the node density of the mesh, appropriate element type, and accurate application of boundary conditions. However, these factors form a trade-off between accuracy and simulation time. Therefore, extra precaution is needed in meshing the geometry, especially during the analysis of the effect of special computational techniques such as fluid–structure interaction (FSI).

3.1.3 Biomedical MEMS Micropump

The conceptualisation, development and realisation of MEMS-based micropumps are intensely supported by modelling and simulation platforms that are based on

Heart valve design

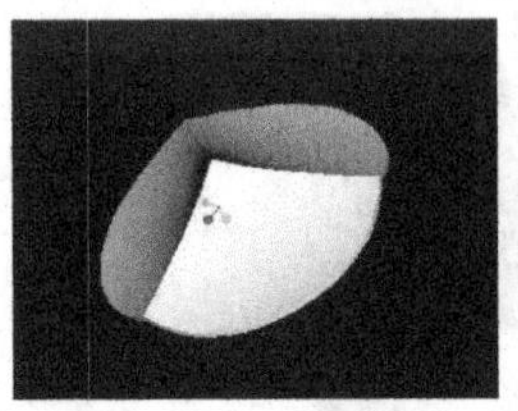

Valve mold design

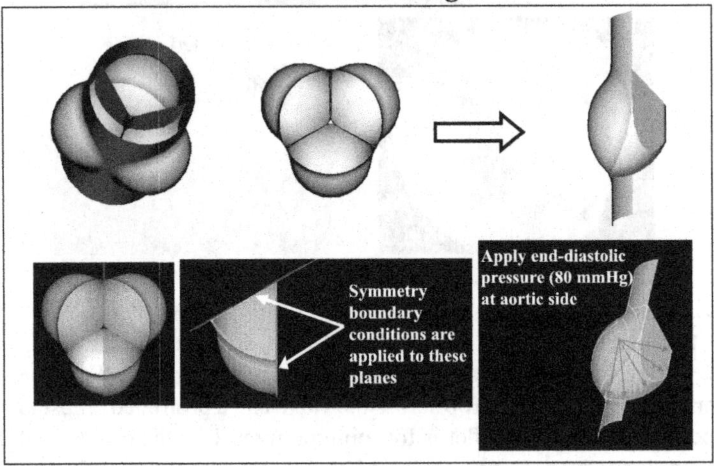

FIGURE 3.2

Computational design analysis of a tissue heart valve and valve mould design. Computational analysis can be used to complement physical prototyping and design analysis of the heart valve mould and the tissue valve. The use of CAD and CFD software can assist in realising the design configurations of the heart valve implant and mould based on reliable scientific principles. This can reduce cost and time in manufacturing the real prototype used in clinical testing. Computational design analysis is a useful approach for this type of biomedical product development as demonstrated by the CAD models of the heart valves and mould. (Courtesy of Prof. Joon Hock Yeo, Nanyang Technological University, Singapore.)

CAD and computer-aided engineering (CAE) tools, which are combined with numerical simulation tools. Such tools substitute the traditional approaches that were based on experimental investigations with several design and fabrication cycles until the optimal specifications are satisfied. Numerical simulation tools have become an essential key for designing and manufacturing complex micropumps and microvalves

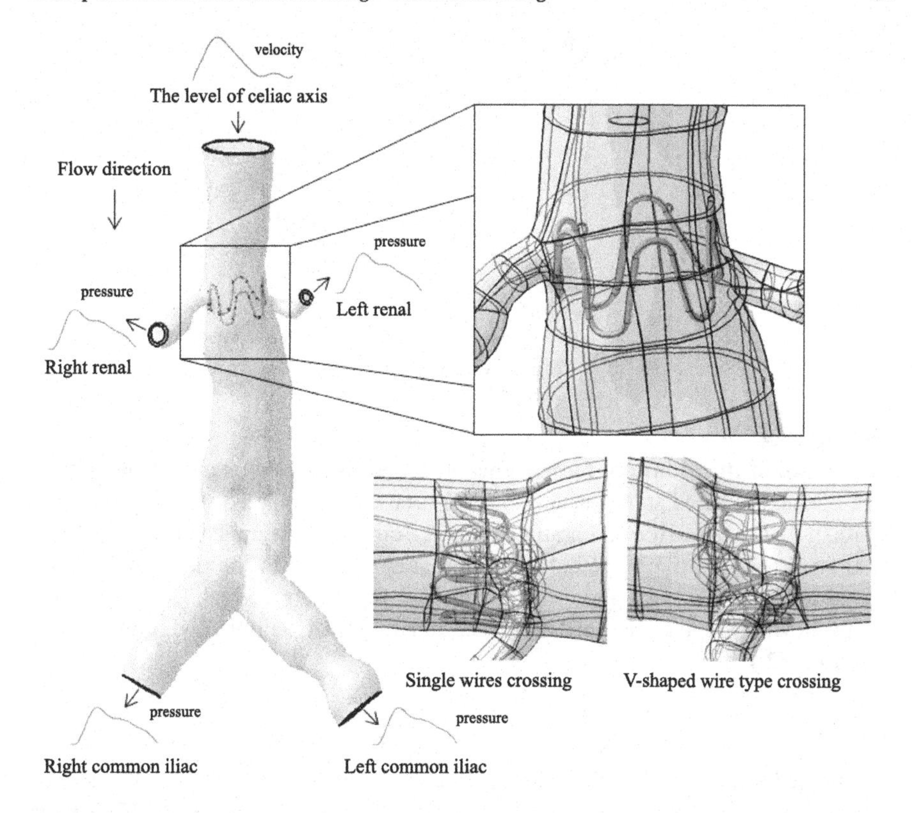

FIGURE 3.3
Simulation of suprarenal stents. Inside the abdominal aorta with stent wires crossing the renal artery ostia. Normal flow velocity is applied to both main abdominal aorta and side branches (renal arteries and common iliac arteries) to demonstrate the interference of stent wires with renal blood flow.

with higher performance and reliability, reduced costs, shorter development cycles and time-saving approaches, compared to traditional approaches.

A major technical barrier in the development of micropumps for biomedical applications lies in the lack of understanding of physical phenomena. Adding to the complexity, the functionality of most of the MEMS-based micropumps made of interactions between multiple physical fields, and relies on thermal, mechanical, electrical, magnetic and/or fluidic field interactions in performing their intended functions. Therefore, further understanding in multi-field microsystems is highly regarded in developing micropumps for critical biomedical applications.

Furthermore, to support the successful implementation of MEMS-based drug delivery systems, a spectrum of numerical simulation tools are needed, such that

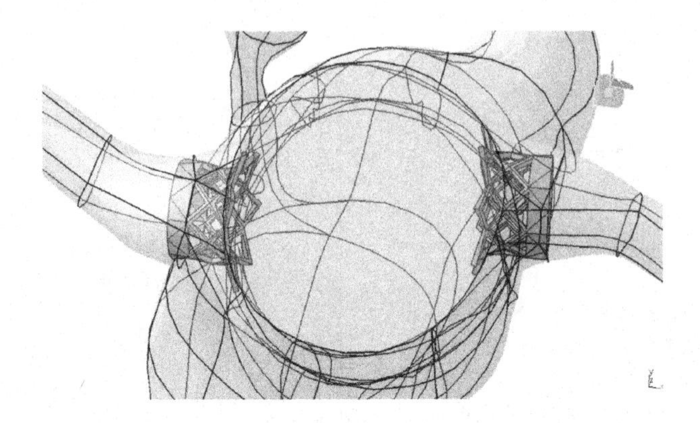

FIGURE 3.4
Simulation of intraluminal appearance of fenestrated renal stents. The appearance of the simulated stent inside the renal arteries is simulated with a protruding length of 5–7 mm into the abdominal aorta. The circular appearance of the simulated renal stent is based on the actual intraluminal appearance of the metal wires after being implanted inside the abdominal aorta. This represents the realistic patient situation following implantation of fenestrated renal stents. (Reprinted with permission from Ref. 52.)

the coupled effects of different physical fields could be analysed simultaneously. In recent times, an increasing trend is identified in the use of FEM and FEA tools for Bio-MEMS design and simulations. Similarly, FEA-based CFD is becoming a prominent numerical method in microfluidic flow analysis problems.

Here in this book, an approach is presented on developing advanced modelling and analysis capabilities for low-powered micropumps, especially for an implantable, battery-less, fully passive device that is targeted at micro-drug delivery applications. The new 3D modelling and simulation methodology is a combination of 3D multi-field analysis, and multiple code coupling capabilities in commercially available modelling and simulation tools ANSYS and CFX. The generalised simulation approach is such that the ANSYS-based FEA code and CFX-based CFD code are simultaneously executed during the proposed multi-field analysis in order to successfully simulate the flow characteristics of the valveless micropump.

Such micropumps can be designed to be interrogated using conventional RFID technology, where low-power circuits can obtain their power from an electromagnetic field.[53,54] As depicted in Fig. 3.5, the micropump structure is mounted on top of a output inter-digital transducer (IDT) of a SAW device. A SAW device is a passive component that consists of a piezoelectric substrate, input IDT and output IDT. The input IDT is connected to a micro-antenna for wireless communication.

The micropump structure consists of a thin conductive microdiaphragm, a pumping chamber and an inlet and outlet made of flat-walled diffusers.

In this chapter, the advanced modelling and analysis of a full 3D model of the micropump is presented as an effective methodology to be utilised at the design stage. Furthermore, FEA-based CFD modelling and simulation results are presented in order to estimate the micropump performance under a low actuation signal and also to investigate the low-powered operation of the device.

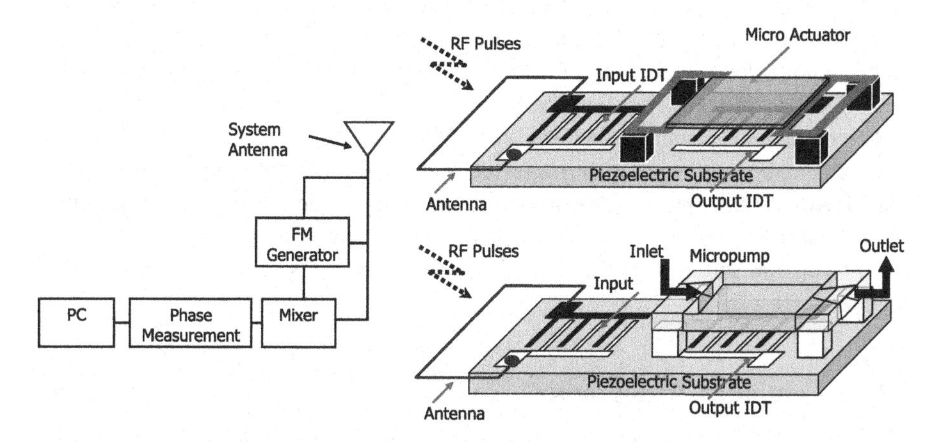

FIGURE 3.5
A conceptual view of a SAW-based interrogation of a 3D micropump (or micro-actuator). The micropump consists of inlet and outlet diffuser elements for flow rectification, the pumping chamber and the microdiaphragm. The pumping chamber with the microdiaphragm is placed on top of the output inter-digital transducer (IDT) of the SAW device separated by a thin air gap. The SAW device consists of a piezoelectric substrate, input IDT and output IDT. The input IDT is connected to a micro-antenna for wireless communication. A SAW correlator can be used for secure actuation.

3.2 Medical Imaging and Reconstruction

Various medical imaging modalities exist. The creation of medical images using X-rays or CT, ultrasound MRI, positron emission tomography (PET) or single photon emission CT (SPECT) can be used to reconstruct 3D models of the pathologies.[55] Different scanning modalities have different sensitivities to the pathological properties. For example, MRI is suitable for imaging cardiac structures as it provides good contrast of the tissue and blood, whereas CT is excellent for providing the geometrical data of bone structures such as the skull.[56] PET or SPECT are able to detect the tumours within the brain, but provides poor registration of the skull.[57] Very often, the rendering software is able to perform spatial and temporal visualisation of these models effectively.

Due to the relatively immense diversification of the medical imaging scanning modalities and protocols, we shall limit our scope to present only MRI and CT, which are widely used for anatomical reconstruction and medical image visualisation. The medical imaging techniques of MRI and CT are used in reconstruction of organs and tissue layout. In addition to anatomical structures, it is worthwhile noting that VIE is a technique that is based on CT and is able to extract geometrical information of biomedical devices such as stent grafts. Therefore, based on these medical imaging and computational visualisation techniques, MRI and CT are typically used to obtain cardiac chamber construction, heart valves, stent structures and nasal cavity geometries. Here, we examine the role of medical imaging in measuring the performance of the biomedical device in its anatomical environment. We explore the geometrical information and also the functional flow measurements provided by the image scanners.

3.2.1 Computed Tomography

CT is one of the most popular medical imaging modalities for 3D anatomical modelling. This can be achieved by the measurement of the myocardial geometry followed by medical image reconstruction.[58] The rendering and visualisation of anatomical heart structures can be approached using video or graphics interface standards to enable an effective functional performance. An appropriate graphics controller must be compatible with such graphical interfaces in order to allow quality display of the rendered view.

The reconstruction of the anatomy based on segmented scan slices produces the computational mesh that can be rendered and patched with colours. The display of a 3D model is rendered onto a 2D screen for viewing. There is a need to design good user interface that is linked with hardware manipulation as well as presentation capabilities that will improve the visualisation processes and information display of the rendered objects.

The utilisation of medical image visualisation software assists in surgical planning. In this section, we describe our developed system that can offer detailed and informative access to anatomical details for the pathology. Current practice requires the use of rapid prototyping tools to create physical models of the pathology, but this is sometimes fundamentally limited by the physical nature of the models and lack of flexibility in disassembling the model components. For example, obtaining multiple slices of an anatomy will require various sections and putting them together again at the ease of the user. In contrast, a software-based modelling system is capable of combining virtual reconstructions on the computer with physical display panels to offer unprecedented flexibility in the visualisation of sections of the pathology.

A graphical rendering framework may employ CT scans of a human part to obtain geometric data, which are then used for the visualisation, with specific application in the present study of the structures in the part.[59] For example, computational modelling of the cardiac chamber can be accomplished using time-resolved anatomical slices of the chamber geometry. The structure evolution and physiologically relevant characteristics are examined. The 3D reconstruction of cardiac structures based on CT scans allows us to achieve a higher physiological accuracy of modelling due to excellent tissue and blood intensity contrast. The measurement tool effectively extracts the wall profile of the cardiac structures *in vivo*. Then, we can present the structural characteristics on the computer monitor using video or graphics interface standards. A database of the anatomical structures of the heart that are derived from the CT images of healthy subjects can be utilised for reconstruction of the left ventricular chamber and the aortic heart valve to carry out visualisation and flow analysis.

One other example of the applications of CT reconstruction in cardiology is the examination of exact characteristics of the blood flow prior to entry into the aorta that can be constructed in a simulation platform. Modelling of valve closure and opening will be more realistic by taking into account all the effects of the non-uniformity of the flow. This study is vital because of the importance in establishing a standard for more accurate heart valve testing. The CT reconstruction and the generation of 3D heart valve mesh in the transient modes open up many new opportunities for flow analysis and examination of valve leaflet behaviour. It will be of interest to cardiologists and physiologists to use information on the structural changes of the cardiac chambers to explain flow phenomena in the heart.

In practice, the construction of CT images in a 3D space for specific time frames, and measurement of flow within the structure in the computational model and its representation of the velocity profile using a vector plot can be achieved. The flow visualisation within a heart will serve as a standard to establish computational and experimental test models that are constructed for the testing of heart valves or observation of their operation. The components of rotation and translation of blood, as well as its velocity variation can be reconstructed for the aortic flow in those models. It may also be worthwhile noting that the MRI technique is also capable of extracting the geometrical information of anatomical structure for reconstruction;[60] however, it is less popular than using CT due to poorer tissue–blood intensity contrast.

3.2.2 Virtual Intravascular Endoscopy

Understanding the 3D relationship of aortic branches and intraluminal stents following endovascular repair of AAAs will aid vascular surgeons to evaluate how the stent grafts situate within the aorta after placement. Conventional 2D and 3D CT images lack the intraluminal views required for the assessment of the endovascular stent grafts relative to the aortic artery branches. However, this is easily overcome with VIE as it allows direct visualisation of the aortic artery ostia and endovascular stents. Our previous studies have demonstrated the usefulness of VIE in the follow-up of both suprarenal fixation and fenestrated repair of aortic aneurysms.[3,52] Figure 3.6 is an example showing the VIE visualisation of suprarenal stents crossing the renal artery ostium, while Fig. 3.7 demonstrates the VIE visualisation of the intraluminal appearances of fenestrated renal stents.

3.2.3 CT Reconstruction of the Nasal Cavity, Pharynx and Larynx

In this example, the nasal cavity, pharynx and larynx are presented. Sets of CT scans that included from the top of the nasal cavity down to the larynx were obtained. The images are in DICOM format and therefore need to be viewed with a DICOM reader. Figure 3.8 shows two slices taken from the directory set of images. In the corners of the images are specific details regarding the scan protocol. In the example, a 37-year-old non-smoking, Asian male was scanned using a helical 64-slice multidetector row CT scanner. In DICOM images, data is embedded into the file directory and this is displayed in the slice images. The data reveals details of the scan, such as the image resolution 512 × 512 pixels, the field of view (FOV) 320 mm, the power rating 120 kV peak and 200 mA and the gantry tilt, which is −8.00 mm.

The segmentation of the scanned images is performed through the freeware ITK-SNAP. The CT images are coloured by tissue density represented as grey scale values. In this example, the airway filled with air is identified by selecting an appropriate intensity range. This ensures that a boundary surrounding the air region can be identified. Screenshots showing the segmentation of a nasal cavity are given in Fig. 3.9, which uses the active contour segmentation algorithm. To initiate the active contour segmentation, the region needs to be seeded by manual selection. Successive iterations are performed which grow or "snake" through the geometry, automatically filling up the airway. The segmentation process takes less than 10 min on a desktop computer with 4 Gb RAM, 3 GHz processor and 256 Mb Video RAM. Because the external air surrounding the face is contiguous with the nasal cavity via the nostrils, it will also be selected during the automated region growing segmentation. This region can be included if the simulation is to explore the effects of the external nose and the outside airflow on the inhalation/exhalation flow profiles. On the other hand, it would be deleted out of the geometry if the nasal cavity is the only region required. Recent studies have begun incorporating the external nose and the external airflow to investigate their influence.[61,62] The final exported model was in stereolithography (STL) format which is readable in most CAD software and CFD mesh programs.

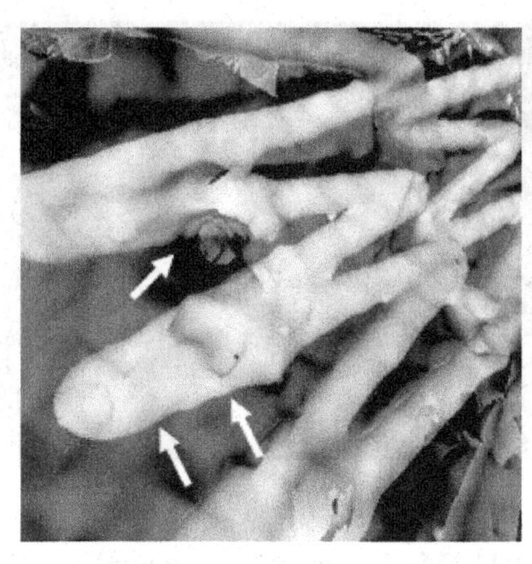
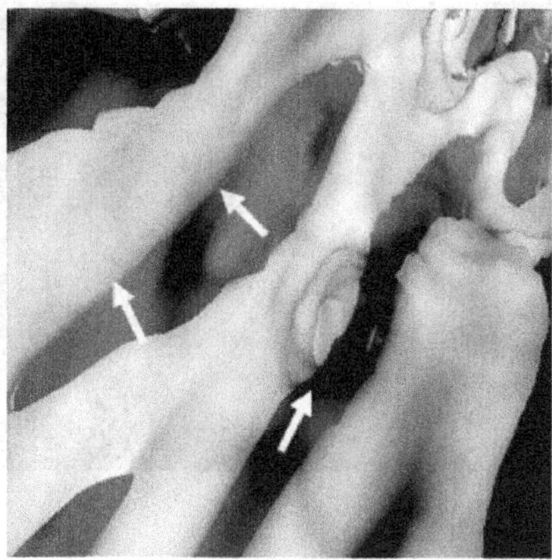

FIGURE 3.6
VIE images of suprarenal stent wires. Stent wires crossing the renal artery peripherally (left) and centrally (right). Short arrows point to the renal artery ostia, while long arrows refer to the stent wires that cross the renal artery ostia. VIE clearly demonstrates the intraluminal relationship between suprarenal stent wires and renal artery ostia.

FIGURE 3.7
VIE views showing the appearance of fenestrated renal stent. Showing the circular appearance of fenestrated renal stent (left) and circular appearance with flaring effect at the lower part of the stent (arrows at right image). The accurate evaluation of intraluminal appearance of the fenestrated renal stents can be clearly obtained with VIE images, while traditional external visualisations fail to provide these details. (Reprinted with permission from Ref. 3.)

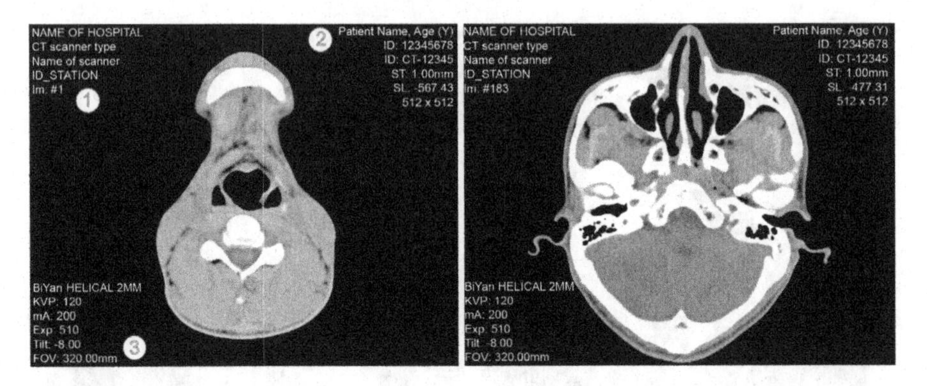

FIGURE 3.8
CT scans viewed in 2D using the free DICOM viewer "MicroDicom". In the images specific details on the scan protocol are displayed and labelled as 1, 2 and 3 in the figure. Data in these section are: 1 — hospital, scanner details and scan image number; 2 — patient details and image resolution in pixels (512 × 512); and 3 — type of CT scan (helical in the example) and scan protocol.

3.2.4 Magnetic Resonance Imaging

While medical imaging is typically used to visualise the geometrical structure of the tissue or organ, functional medical imaging can obtain specific quantities of the organ functions. MRI using the phase contrast protocol has been applied to measure flow in the cardiovascular system.[5] Other examples include diffusion MRI which is another imaging protocol that measures the local micro-structural characteristics of water diffusion within tissues.[64]

The harnessing of MRI for the quantification of flow in the cardiovascular system has been studied.[65] Velocity-encoded cine MRI (VEC-MRI) is a new method for characterising flow patterns in the heart.[66] An improved technique of determining the nature of flow in the left ventricle as well as the transient flow variation of blood through an aortic heart valve based on CT/MRI images is proposed. Using the measured data, we aim to develop a "gold standard" for heart valve testing in particular to the quantification of the (1) variation of velocity with respect to time for blood upstream of the aortic heart valve, (2) cross-sectional flow profile of blood across the aorta and (3) rotational characteristics of blood as it flows out of the ventricle and through the aorta.

The determination of the flow characteristics based on the *in-vitro* retrieval of information from the human body through CT and MRI, and the *in-vivo* measurements of flow based on imaging of the heart reinforce the reliability of determining the pulsating flow waveform using velocity as the quantification of the flow. This waveform

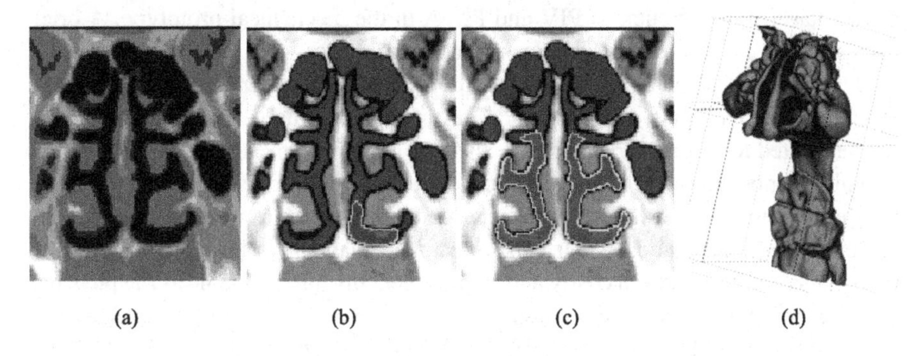

(a) (b) (c) (d)

FIGURE 3.9
The 3D active contour (snake) segmentation of a human nasal cavity using the open source application "ITK-SNAP".[63] (a) Unprocessed DICOM image. (b) Intensity thresholding creates the outlined blue region, and the active contour segmentation (maroon colour) after 100 iterations shows the selection "snaking" through the region. (c) Active contour segmentation after 400 iterations. (d) Final output model as an .stl file.

is further verified by testing it on an artificial circulatory system and comparing the simulated pressure waveform given by the mechanical test rig with the physiological pressure waveform of the real heart. The research comprises the experimental and verification techniques that will be described in greater detail.

For the *in-vitro* retrieval of flow characteristics, the change in the volume of the ventricle at each time frame determines the volumetric flow rate through the heart valve over a period of time. Reconstruction of the image slices and segmentation of the ventricle in the 3D rendered volume enables the calculation of the capacity of the chamber at each time step. The variation of capacity causes the change in the volume of blood inside the ventricle and in effect, determines the volumetric flow rate of the blood through the aortic valve. The dilation of the elastic aorta causes a change in the area of the flow through the valve. The division of the volumetric flow rate and the area change with respect to time gives a good estimation of the velocity of the flow through the valve.

3.3 Mechanical Prototyping

Experimental results are a necessity for verification of any CFD framework. The physical model of the human anatomical part is first manufactured by rapid prototyping techniques. Velocity profiles and particles analysis are then evaluated through *in-vitro* laser imaging such as PIV and PDPA in the mechanical prototype. A brief discussion on each process is discussed herein.

During surgical planning, the fabrication of pathology for medical experts to reference various points of intervention has been widely used. The pathology may not be limited to cardiac structures but also any organ or bone structure in the body. Usually, computational reconstruction can be achieved after effective segmentation for the region of interest. The model can be passed to rapid prototyping software to develop physical models that can be handled manually. However, the cost of such manufacturing is often costly and environmentally unsustainable as the physical model cannot be recycled. In the computer environment, modelling is versatile and display entities can be created or deleted efficiently and effectively.

3.3.1 Rapid Prototyping by Stereolithography

Physical modelling of human anatomical parts is made possible through biomedical imaging (CT and MRI scans), anatomical reconstruction and computational modelling (CAD models), and then mechanical construction by a technique known as rapid prototyping. Rapid prototyping is also known as solid freeform fabrication. It is a process whereby creation of models take place physically via an *additive* or *subtraction* process based on a part file in the database. By definition, an additive process creates a model by layering materials, while the subtraction process subtracts material from a physical block to create the same model. This manufacturing technology has been applied in medicine and biomedical engineering since 1990.[67] It is being reported that at present, rapid prototyping technology has been widely used in neurosurgery, oral and maxillofacial surgery, orthopaedics and tissue engineering.[68,69] Application in vascular reconstruction has also already been reported.[70,71] With the advancements in medical imaging and image processing, the axial images of arterial vessels and heart structures can be reconstructed and exported in a digital format. Consequently, rapid prototyping technology can change this digital format into a physical replicate for flow experimentation and analysis.

STL belongs to a class of rapid prototyping where a design is created from a CAD model and converted into a compatible STL format. For the STL format, the surfaces of the solid part is formed by triangulation. A physical model can be produced through stereolithography based on construction of a 3D stereo printer lithograph.[1] This technique is based on the additive process. The stereo printer uses an ultraviolet laser beam to build 2D layers by curing the photopolymeric resin (in the form of liquid or powder) that is deposited onto the prototyping platform (Fig. 3.10). This technology is known as laser additive manufacturing (LAM). Its advantages include unrestricted geometrical and material fabrication.

In LAM, an ultraviolet laser traces out a 2D cross-section slice of a vat of a photosensitive polymer resin. Polymerisation occurs as the laser traces the 2D slice

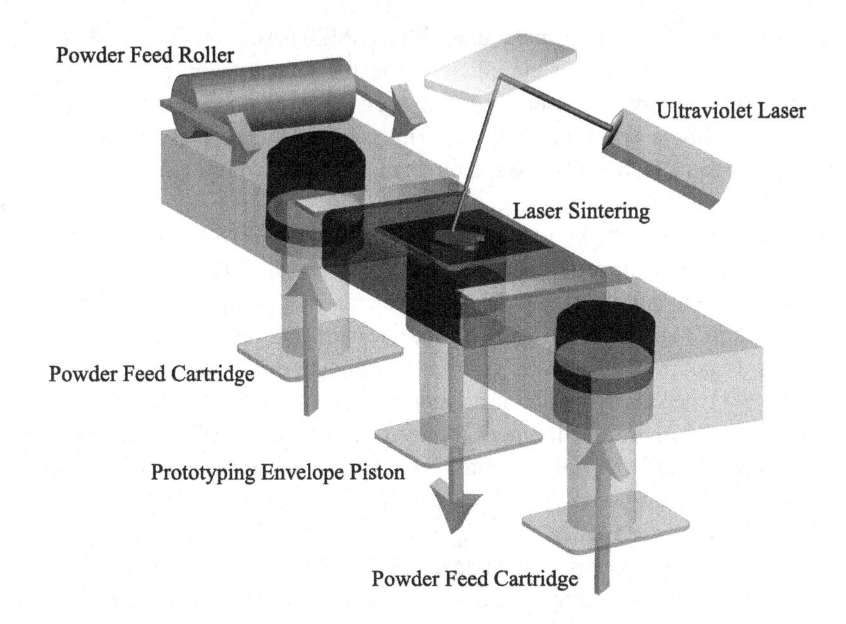

FIGURE 3.10
Rapid prototyping by STL. The laser beam moves according to the input data and traces out a 2D slice. The laser polymerises the resin and creates a solid. In this case, resin in the form of powder is illustrated. Once the layer is complete, the platform is lowered and the process repeats until the whole object has been reconstructed.

while the excess areas are left as liquid or powder (depending on the raw material used). Upon completion, the solid cross-section is lowered by a distance equal to the layer thickness known as the "build layer thickness" that is typically at 0.075 to 0.05 mm. This continues until the entire replicate is created. Manipulating the laser to sweep across the polymer resin requires dedicated control to ensure that the resin can cure while the laser beam spot is traversing at a sufficiently high speed, so as to minimise the time for the model formation process. In general, a higher laser traversing speed may lead to a lower power consumption, which relates to a lower manufacturing cost. However, a high layering rate also leads to lower accuracy and durability of the formed part. The prototyping envelope piston then elevates the solid formed part out of the liquid resin to be cleansed with a solvent. The final stage is to complete the cure of the photopolymer part in an ultraviolet oven.

Up to today, advances in the rapid prototyping technology have given rise to new processes that deal with plastic, nylon, polycarbonate or even metal powders such as aluminium or steel. For instance, the selective laser sintering process fuses such powders using heat from a precise guided laser beam. In particular, selective laser melting (SLM) and laser metal deposition (LMD) pertains to the LAM process, where the key challenge is in the efficient transfer of heat from the laser into the material powders. A higher laser traversing speed (e.g., 500 m/min) and deposition rate corresponds to a faster build up rate. This is also dependent on the laser power that can go up to 1 kW. In most cases, their "build layer thickness" ranges from 10 μm to 0.3 mm.

Another innovation is the fused deposition modelling (FDM), where the part is developed by layering materials ejected from a hot-glue gun. Other processes include laminated object manufacturing (LOM), which creates the part from layering papers and fusing them together. The last two processes are generally cheaper and safer to use as they do not involve metallic powders and high laser power.

Figure 3.11 depicts a human aorta model manufactured with LAM. It models the physiological environment where the medical device resides, which serves the propose of producing a visual aid, and device testing. It can be noted that from the 3D surface shaded image shown in Fig. 3.11, the aortic branches such as renal arteries are clearly demonstrated. Then, utilising advanced image processing techniques, the computer-modelled prototype can be developed. It should be noted that the external surface of the phantom is rough and irregular as we only require the internal surface to be as smooth as possible. Figure 3.11 also shows a human aorta phantom that is built with rapid medical prototyping and clearly indicating the fusion line in the middle of the phantom.

In general, rapid prototyping by the aforementioned techniques can give rise to the effective communication of complex or constantly modified design parts. It can also be used as a functional design prototype for marketing and as a form of preliminary design evaluation, or even as patterns for prototype tooling. They are becoming increasing popular among organisations dealing with product development and manufacturing.

3.3.2 Technical Limitations

While STL offers an exciting and quick process to manufacture simple and complex parts that may never have been achievable in the past, the process still requires some care and preparation. Firstly, the fidelity of the final product is highly dependent on the computational STL file that is fed into the STL machine. The STL data is often converted from CAD software which in turn may have come from rebuilding the computational model from MRI or CT scanned data. For example, in the manufacture of a complex geometry such as the nasal cavity or lung bifurcations, the initial data is obtained from CT or MRI scans that take 2D contiguous "slices". From these scans, the point clouds and then interpolation of surfaces across the 2D slices are performed. Thus, the final STL file becomes dependent on the initial 2D slice spacing because of the interpolation process.

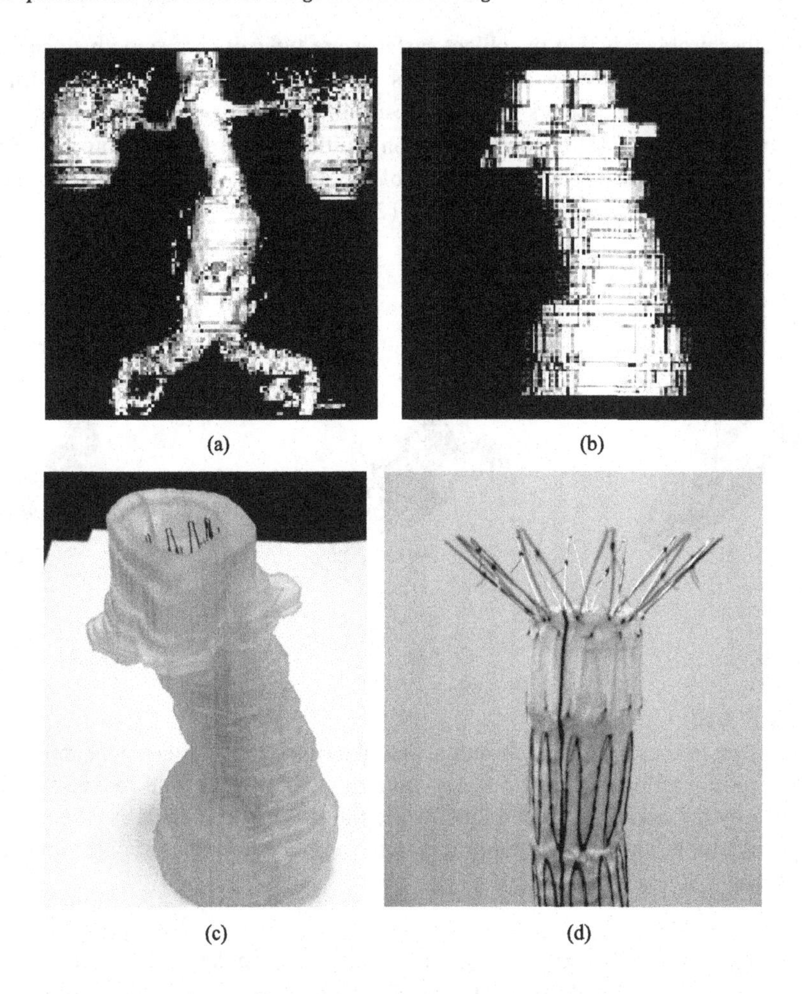

(a) (b)

(c) (d)

FIGURE 3.11
Human aorta model produced by rapid prototyping. (a) A 3D surface shaded image of an AAA selected for phantom design and construction. The aortic branches such as renal arteries are demonstrated clearly. (b) The computer model prototype after image processing and editing. It is observed that the external surface of the phantom is rough and irregular as we only require the internal surface to be as smooth as possible. (c) A human aorta phantom built with rapid medical prototyping. Normal aortic branches and cephalad of the aneurysm can be clearly seen on the image. A commercially available aortic stent graft (d) is placed inside the phantom to simulate the normal endovascular repair of aortic aneurysm. With the stent graft in place, the phantom can be used for experiments to optimise CT scanning protocols and evaluate image quality.

Another problem is that the object surfaces are typically not smooth since STL processes approximate curved surfaces as stair-stepped layers of discrete thickness.[72–74] Thicker build layers generally result in greater airway surface roughness in the build direction.[1] These surface discontinuities need to be accounted for. One method is to consider which 3D projection of the object that the slices should be built upon (coronal, axial or saggital; see Fig. 3.12 for orientations).

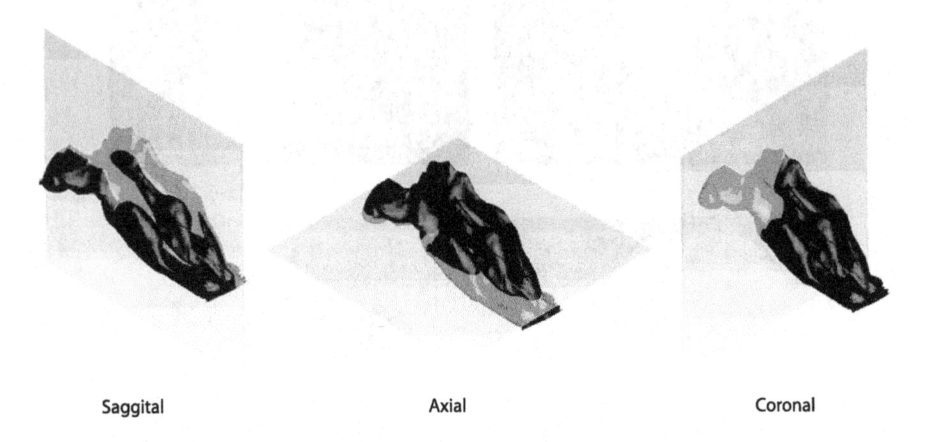

FIGURE 3.12
Orientation planes used in anatomical visualisation. Three-dimensional presentations of cross-sectional slices from a volumetric object can be selected. Axial slice is shown by the horizontal slice plane, while the two vertical slices are the coronal (perpendicular to anterior–posterior axis) and saggital slices that are orthogonal to each other.

Other considerations include the time it takes to complete the manufacture. A typical run might take 6–12 h, while building large objects (up to 25 cm in the *x*–, *y*–, and *z*– directions) may take several days. Although this contradicts the term "rapid prototyping", the complex parts may not be achievable in any other way. In addition, STL is an expensive process. Machines and the polymer resin typically cost a lot. High levels of ventilation are also needed to extract the fumes created during the polymerisation step.

3.4 Summary

In general, the design of a biomedical device is based on a set of integrated preliminary concepts structured on its intent, its functionality and its safety or aesthetics.

The conceptual modelling of the biomedical device can be achieved by CAD, and the physiological surroundings can be constructed by medical imaging, all within a computational environment. Mechanically, rapid prototyping can be used to construct preliminary versions of the biomedical device or its anatomical environment. The designs can be imported into a computational modelling platform for the simulation of its haemodynamics properties. Physical prototypes can be tested for their design effectiveness by mechanical simulation in a test rig.

In the next few chapters, we will provide the fundamental concepts related to biomechanics, medical imaging, computational modelling, biofluid analysis and also the integrated techniques of producing a concise biomedical device. Based on generic road maps that pertains to its development, we synthesise the scientific concepts and present biomedical device research in a systematic and generic manner.

3.5 Questions

1) Name two different types of additive rapid prototyping techniques.
2) Name the different types of medical imaging reconstruction techniques.
3) Give two examples of using MRI to obtain specific quantities of the organ functions.
4) In the STL format, how do surfaces of solid parts form into the STL format?
5) What are some of the technical limitations of the STL process?

4

Medical Imaging and Visualisation

CONTENTS

4.1 Computed Tomography 37
4.2 Virtual Intravascular Endoscopy 38
4.2.1 Generation and Presentation of VIE 40
4.2.2 Generation of VIE Images 42
4.2.3 Threshold Range Along the Abdominal Aorta 42
4.2.4 Optimal Threshold Selection 43
4.2.5 Generation of VIE Images with Aortic Stent and Artery Lumen Together 44
4.2.6 Aortic Stent Wire Thickness on VIE Images 47
4.2.7 Image Display and Interpretation 47
4.3 Optimal CT Scanning Protocols for VIE Visualisation 51
4.4 Summary 54
4.5 Questions 54

4.1 Computed Tomography

CT is a digital imaging technique that produces sectional or axial slices of the body without any intersection interference or blurring. The method was first developed in a commercial X-ray machine by Godfrey Hounsfield from the UK in 1973. It was immediately recognised as a useful diagnostic imaging technique as it could enhance smaller contrast difference. Thus, it is able to differentiate between normal and abnormal structures. Since then, CT has revolutionised not only the diagnostic radiology, but also the entire field of medicine.[75]

The introduction of spiral CT in the late 1980s represented a fundamental evolutionary step in the development and continuing refinement of CT imaging techniques.[75,76] Until then, the examination volume had to be covered by sequential axial scans in a "step-and-shoot" mode. Axial scanning requires long examination times because of the interscan delays necessary to move the table incrementally from one scan position to the next and unwind the cable, thus it is prone to misregistration or loss of anatomical details due to potential movement of the relevant anatomical structures between two scans (by patient breathing, motion or swallowing). Besides, only a few slices are scanned during maximum contrast enhancement when the contrast medium is used. These problems may be overcome if the scan speed is increased and interscan delay is eliminated. The development of spiral CT possesses these features which help to overcome the above problems.

In 1989, spiral CT became available as a mode for continuous volume scanning.[75] With spiral CT, the patient table is continuously moved and translated through the gantry while scan data are acquired simultaneously. A prerequisite for spiral CT scanning is the introduction of slip-ring technology, which eliminates the need to rewind the cable after each rotation and enables continuous data acquisition during multiple rotations.

The purpose of the slip-ring is to allow the X-ray tube and detectors to rotate continuously so that a volume data of the region of interest, rather than a single slice, can be acquired very quickly in a single breath hold. Spiral CT scanning does not suffer from the danger of misregistration or loss of anatomic details. Images could be reconstructed at any position along the patient longitudinal axis, and overlapping image reconstruction could be generated (normally 50% overlap) to improve longitudinal resolution. Acquisition of volume data has become the very basis for applications such as CT angiography.

A significant advantage and development of spiral CT data acquisition is its application in 3D imaging of the vascular structures with an intravenous injection of the contrast medium. This application is referred to as CT angiography (CTA). CTA is achieved with two essential requirements: volumetric data must be acquired with spiral CT scanning, while in the meantime, contrast medium is delivered intravenously. CTA images can be captured when vessels are fully opacified to demonstrate either arterial or venous phase enhancement through the acquisition of both datasets (arterial or venous phase depending on the scan delay). CTA has been widely applied to a number of examinations investigating vascular anatomy and diseases, and it has been regarded as one of the most valuable applications in CT imaging. CTA produces angiography-like images non-invasively in a 3D format, thus it has replaced conventional angiography in many applications, such as imaging thoracic and abdominal aorta, pulmonary and renal arteries (Fig. 4.1(a)). Also, CTA provides valuable information about aortic stent grafts in relation to the artery branches (Fig. 4.1(b)).

4.2 Virtual Intravascular Endoscopy

VE or computed endoscopy is a new method of image data visualisation using computer processing of 3D image datasets (such as CT or magnetic resonance scans). It can provide visualisations of specific organs similar or equivalent to those produced by standard endoscopic procedures. Conventional CT and magnetic resonance scans acquire cross-sectional "slices" of the body that are viewed sequentially by radiolo-

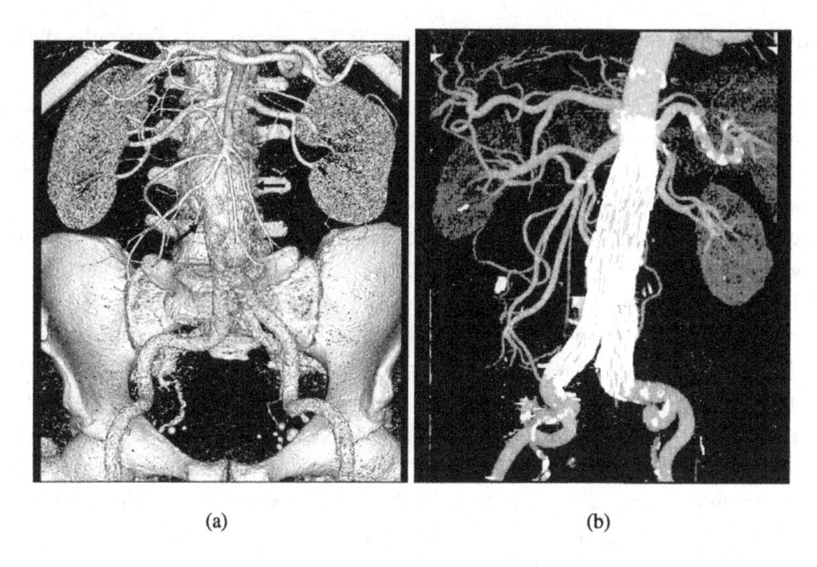

(a) (b)

FIGURE 4.1
CTA of an aortic aneurysm. CTA-generated 3D visualisation of an infrarenal aortic aneurysm (arrows in (a)) is clearly displayed. A high-density stent graft is also visualised on the 3D CTA image together with the abdominal aorta and its artery branches (b).

gists who must imagine or extrapolate from these views what the actual 3D anatomy should be. By using sophisticated algorithms and high performance computing, these cross-sections may be rendered as direct 3D representations of the human anatomy.

Thousands of endoscopic procedures are performed each year. They are invasive and often uncomfortable for patients. They sometimes have serious side effects such as perforation, infection and haemorrhage. VE visualisation avoids the risks associated with real endoscopy and when used prior to performing an actual endoscopic exam, it can minimise procedural difficulties and decrease the rate of morbidity, especially for endoscopists in training. Additionally, there are many body regions not accessible or incompatible with real endoscopy that can be explored with VE. Eventually, when refined, VE may replace many forms of real endoscopy.

Several important body systems may need an endoscopic procedure with invasive probes including the heart, spinal canal, inner ear (cochlea, semicircular canals, etc.), biliary and pancreatic ducts and large blood vessels. These are important anatomical structures ideally suited for VE. VE in the evaluation of colonic polyps (virtual colonoscopy) and bronchial disease (virtual bronchoscopy) has been widely studied in the literature;[77,78] however, the application of VE in blood vessels, namely VIE, is an emerging area and will play a potential role in the assessment of vascular disease as well as surgical treatment follow-up.

VIE is a technique that has the ability to visualise anatomy and potentially pathological changes directly within vessels as if a catheter containing a camera were placed inside the patient's vessel (Fig. 4.2). An intravascular pathologic process such as a thrombus, a tumor or a pulmonary embolus or a plaque, will be visible in the final image (Fig. 4.3). This is different from traditional extraluminal views as VIE allows the evaluation of the structures or pathological changes inside the blood vessels, thus additional information can be acquired to assist further analysis or diagnosis.

Early studies of VIE in aortic stent grafting have confirmed the technique to be a promising one as it can provide intraluminal information about the aortic aneurysm and stent graft which cannot be obtained by conventional methods.[79,80] The ability to obtain objective 3D reconstructions of an internal view of the vessel, its walls and the stent would facilitate both qualitative and quantitative analysis and potentially enhance diagnostic capabilities. Plaques and stenoses are easily identified and can be quantified exactly using the cross-sectional area. The stent structures are clearly visible and the stent position within the vessel and relative to the lesion can be assessed.[80] CT-based VE represents an alternative or addition to conventional angiography.

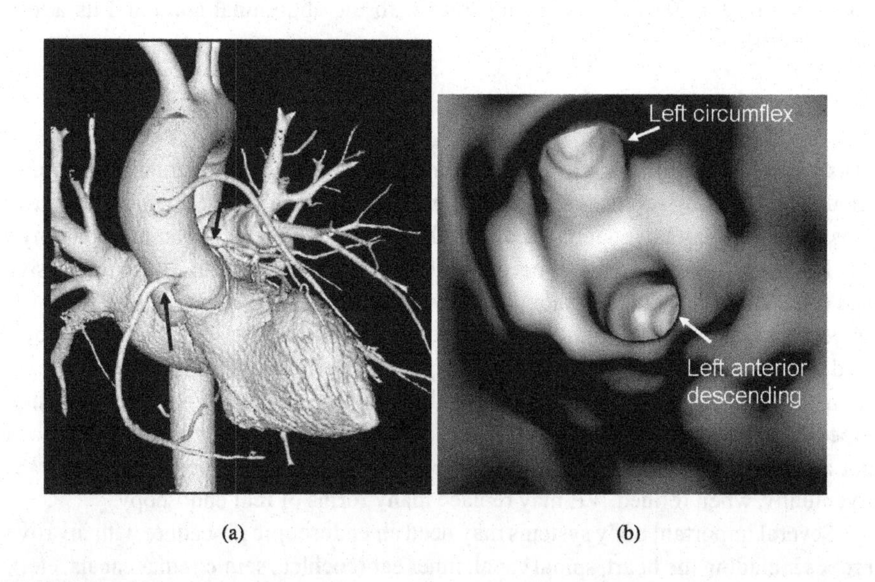

FIGURE 4.2
VIE visualisation of coronary artery ostium. Three-dimensional CTA demonstrates the right (long arrow in (a)) and left (short arrow in (b)) coronary arteries. Corresponding VIE shows clearly the intraluminal view of the left coronary ostia.

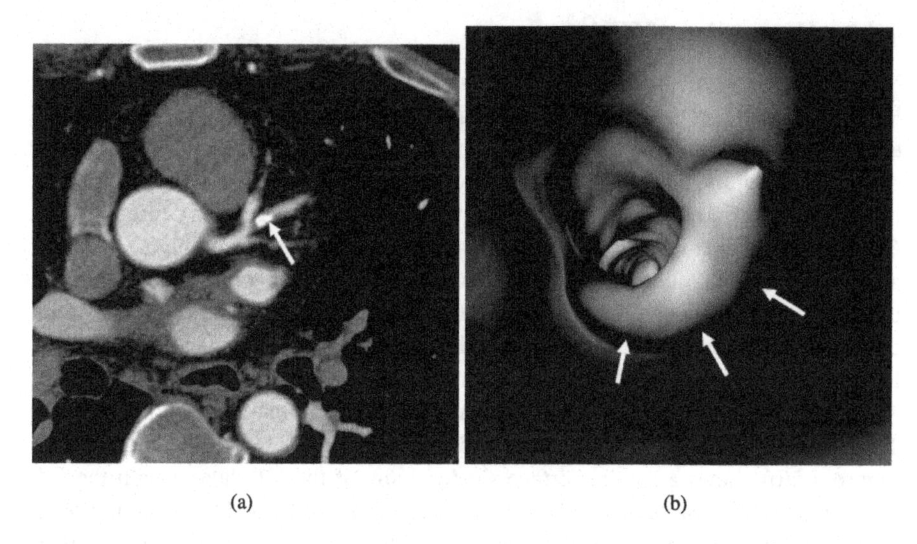

(a) (b)

FIGURE 4.3
VIE visualisation of a coronary plaque. Two-dimensional axial CT image shows a calcified plaque (arrow in (a)) at left coronary artery. Corresponding VIE demonstrates the intraluminal protruding sign of the plaque at the coronary artery wall (arrows in (b)).

4.2.1 Generation and Presentation of VIE

Spiral CT has dramatically improved the performance of CT by converting a 2D modality into true 3D imaging, thus enabling the development of new applications involving volumetric imaging, such as CTA.[81] In turn, CTA has been complemented by a parallel development of postprocessing methods to create a 3D representation of anatomical structures.[80,82] Among these methods, VE is the most recent and is currently under clinical investigation, primarily in the fields of colonoscopy and bronchoscopy.[83,84] A few studies have applied this technique to vascular diseases.

VIE can be generated using data obtained with CTA, magnetic resonance angiography (MRA) or digital subtraction angiography (DSA).[85–89] Although 3D DSA was reported to be useful in creating intraluminal views of blood vessels and stent images, it is an invasive technique and associated with procedure-related complications. Moreover, stent-induced artefacts and safety concerns have prevented MRA from being widely used.[89] Thus, this makes 3D CTA a routine technique in the preoperative evaluation of candidates for endovascular AAA repair and for the postoperative follow-up of stent grafts.

4.2.2 Generation of VIE Images

Virtual bronchoscopy or colonoscopy uses the natural difference between the intraluminal gas and soft tissue and makes rendering of 3D endoscopic images relatively easy.[83,84] In VIE, since blood vessels (arteries) are opacified with the administration of contrast medium, therefore, the contrast enhanced blood needs to be removed from the artery before performing intraluminal visualisation and virtual fly-through. The CT data was prepared for VIE by removing the contrast-enhanced blood from the aorta using a CT number thresholding technique. A CT number range was identified using region of interest (ROI) measurements. Figure 4.4 shows the three locations chosen to be representative of the CT number range of contrast enhanced blood, i.e., the renal artery, aneurysm body and common iliac artery locations. Figure 4.5(a) is a graph of the average CT number at the three ROI locations as shown in Fig. 4.1. These three measurements were averaged to produce the applied threshold value. Figure 4.5(b) shows a caudal surface shaded view of the CT dataset with the CT number threshold applied. Note that the contrast-enhanced blood has been removed from the major arteries. Also, because of the threshold value chosen, high-density bone information has also been removed.

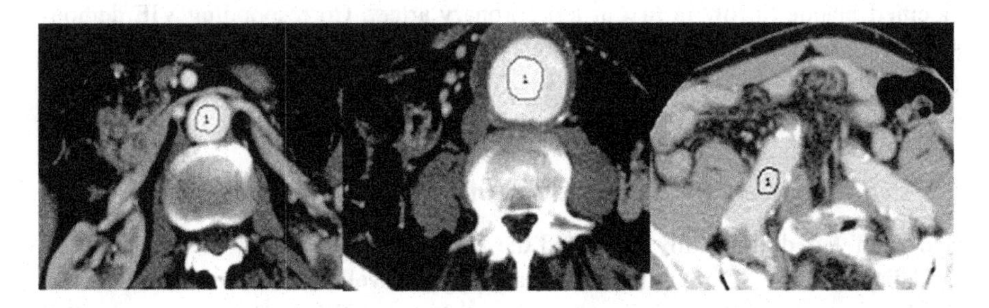

FIGURE 4.4
Measurement of CT attenuation at different anatomic regions. Three anatomic regions of interest are selected at the renal artery, aortic aneurysm and common iliac artery to measure CT attenuation for the generation of intraluminal views.

4.2.3 Threshold Range Along the Abdominal Aorta

An important issue to be considered is variation in the attenuation of contrast-enhanced blood at different levels in the abdominal aorta. This stresses the importance of choosing the appropriate threshold range in different levels of the aorta to obtain optimal VIE images. This is especially apparent in the common iliac arteries, as CT attenuation is sometimes lower in this region than that of the renal artery or aneurysm body during helical CTA. This is usually encountered with early type

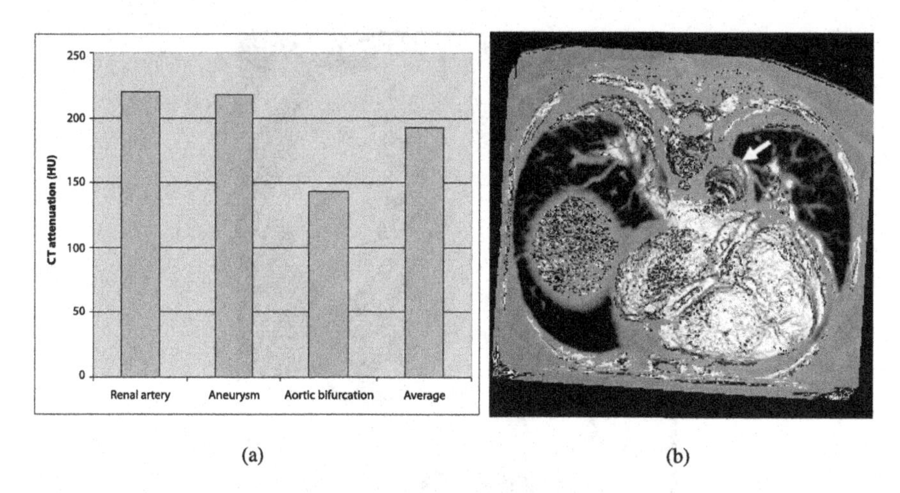

(a) (b)

FIGURE 4.5
Identification of CT attenuation for VIE generation. CT attenuations measured at the above three regions (a) were averaged to allow an endoluminal view of the abdominal aorta. Arrow indicates that contrast-enhanced blood was removed from the aorta after applying the threshold of 180 HU (b).

of CT scanners, such as single-slice CT or four-slice CT scanners. With the current advanced CT models which include 16-, 64- and dual source CT, homogeneous contrast enhancement can be acquired from the coeliac axis to the peripheral artery system down to the distal lower extremity due to a fast scanning speed, ensuring the acquisition of high quality VIE images of the abdominal aorta and its branches. Figure 4.6 demonstrates homogeneous contrast enhancement in a patient with an infrarenal aortic aneurysm with CTA being performed on a 64-slice CT. The CT attenuation in the levels of renal artery, aneurysm body and aortic bifurcation were 276, 290 and 272 HU (Hounsfield unit), respectively. Correspondingly, a threshold of 250 HU was applied to generate intraluminal VIE views of the renal ostium, aortic aneurysm and common iliac arteries, as shown in Fig. 4.7.

4.2.4 Optimal Threshold Selection

The selection of the optimal threshold is important in producing VE images that are free from artefacts. The lower threshold is always set at −1,200 HU to include all soft tissue and vessel lumen. This means that only the upper threshold level has to be changed to remove contrast-enhanced blood. Once an averaged threshold is determined, the upper threshold is progressively changed, in steps of 10 HU to detect alterations in the aortic internal surface. This ensures that floating shape and

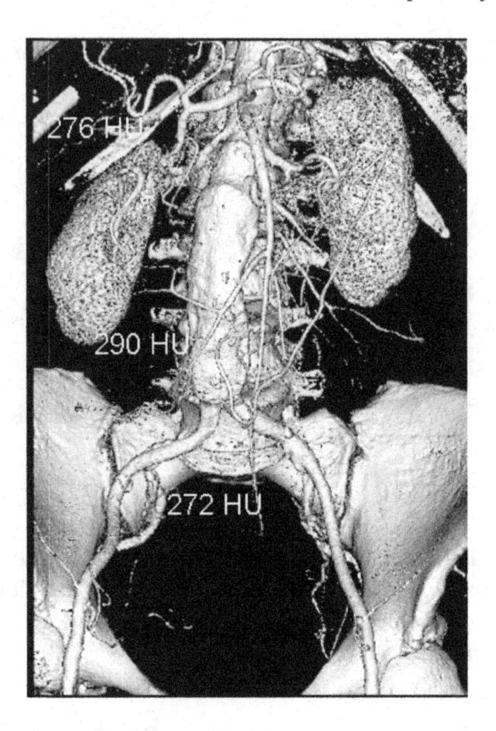

FIGURE 4.6
Homogeneous contrast enhancement in CTA. Consistent contrast enhancement was achieved a patient with abdominal aortic aneurysm undergoing helical CTA on a 64-slice CT scanner. CT attenuation was measured from the level of renal arteries to common iliac arteries with attenuation difference less than 20 HU.

apparent hole artefacts are avoided in the final VIE images. Figure 4.8 shows that the generation of VIE views of right renal ostium is determined by the selection of an appropriate threshold value, as the intraluminal appearance of the renal ostium could be irregular or distorted due to presence of artefacts if an inappropriate threshold is selected, thus affecting visualisation and clinical assessment.

4.2.5 Generation of VIE Images with Aortic Stent and Artery Lumen Together

The purpose of threshold selection is to demonstrate the lumen surface which in effect removes the stent from the image. A procedure is required to visualise both the lumen and stent together to study the stent/ostia relationship. Endovascular aortic stents are composed of stainless steel wires, thus presenting with high-density struc-

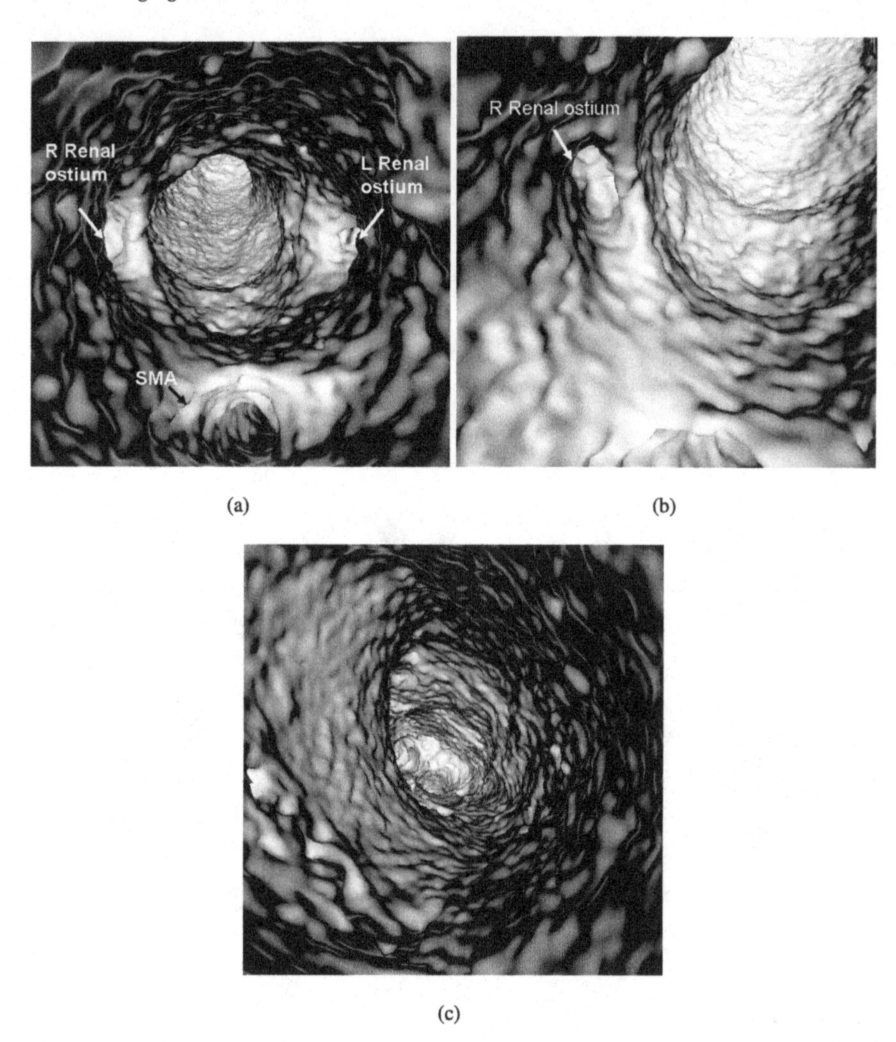

FIGURE 4.7
VIE views of aortic ostia and aneurysm. After applying the measured threshold in Fig. 3(a), which is 270 HU, intraluminal VIE images were generated to look at the aortic ostia (SMA: superior mesenteric artery) (a), close view of the right renal ostium (b), inside the aortic aneurysm (c) looking towards the common iliac arteries.

tures on the cross-sectional views. The CT attenuation of the metal wire is much higher than that of contrast-enhanced blood. Therefore, different threshold ranges are required to generate VIE images of aortic stents and luminal structures separately.

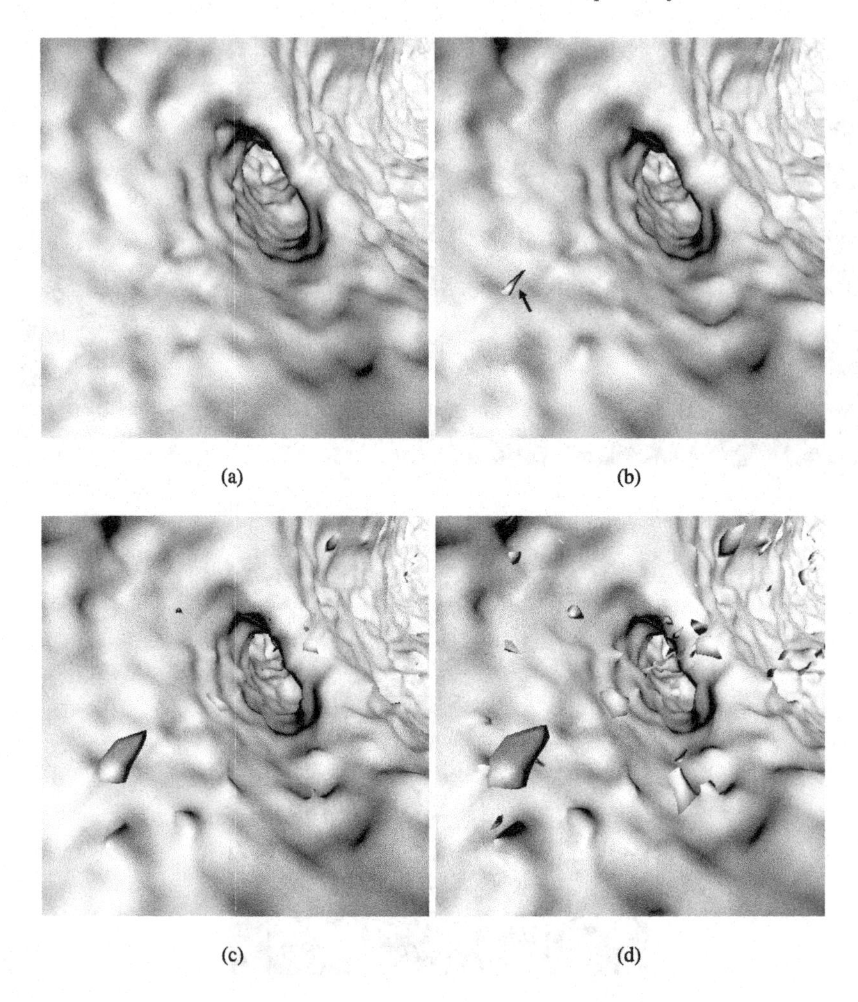

FIGURE 4.8
Effect of threshold selection on aortic ostium visualisation. (a) VIE view of the right renal ostium was clearly displayed when CT threshold of 250 HU was applied. Floating artefacts started to appear when the threshold was increased from 250 HU to 270 HU (arrow indicates the artefacts in (b)), 290 HU (c) and 300 HU (d). Renal ostium becomes irregular or distorted with a threshold more than 290 HU.

Initially, a lower threshold range, e.g., −1,200 to 120 HU, is applied to remove the contrast-enhanced blood and the stents from the aorta leaving only aortic luminal structures such as the arterial ostia and aortic wall intact and unaffected. This VIE image is rendered and saved. Then, a higher threshold range, e.g., 400 to 2,000 HU, is applied to visualise the stent wire only without any luminal or soft tissue structures. This second image is also saved. These two images are created with exactly the same camera and eye coordinates. With the aid of a computer software, the two individual images are added together, which allow the stent wire and lumen to be visualised in a single combined VIE image (Fig. 4.9). Figure 4.9(a) shows the left renal ostium generated with the lower threshold of −1,200 HU and upper threshold of 120 HU to include all the soft tissue and lumen structures and exclude high-density structures such as contrast-enhanced blood and aortic stent. Figure 4.9(b) illustrates the VE image of the aortic stent generated by applying the lower threshold of 500 HU and upper threshold of 2,890 HU to include only the high-density stent metal. Figure 4.9(c) shows the combined VIE image demonstrating the 3D relationship of the stent wire to the renal ostium by adding Figs. 4.9(a) and 4.9(b). In this image, it is clearly apparent that the left renal ostium is covered by two metal wires.

4.2.6 Aortic Stent Wire Thickness on VIE Images

An important aspect to be mentioned in generating VIE images of an aortic stent graft is the appropriate selection of a threshold for displaying the metal wires. If the threshold selected is too low, the stent wire appears to be very thick, leading to an overestimation of the wire thickness. If the threshold selected is too high, the wire appears to be very thin or even disrupted in its structure, thus affecting the visualisation of stent wire appearance and evaluation of its encroachment to the aortic ostium. Figure 4.10 shows examples of the effect of changing the threshold value and wire visualisation. The wire thickness is altered depending on the threshold chosen. An appropriate threshold range is determined by generating VIE images of aortic stents with the wire thickness appearing similar to that seen on axial CT images. From the figure, an example of a stent wire covering a renal ostium can be seen. The left renal ostium appears covered to 50% by the wire. However, we know that the actual wire diameter is 0.4 mm and therefore does not cover the ostium as much as shown in the image. Stent wire thickness is measured on axial CT and VIE and ranges from 1.0 to 2.0 mm. Furthermore, the threshold range is variable at different anatomic locations such as the superior mesenteric artery, renal artery and aortic bifurcation. These threshold ranges are also determined by comparing the wire thicknesses with those on axial CT images.

4.2.7 Image Display and Interpretation

To navigate through a vessel, a series of endoscopic views are generated at intervals along the vessel. Endoscopic views can be presented by using either the fly-through or step-by-step mode (Fig. 4.11). With the former, a number of camera positions are established along a fly path of the aorta and/or its branches and the computer

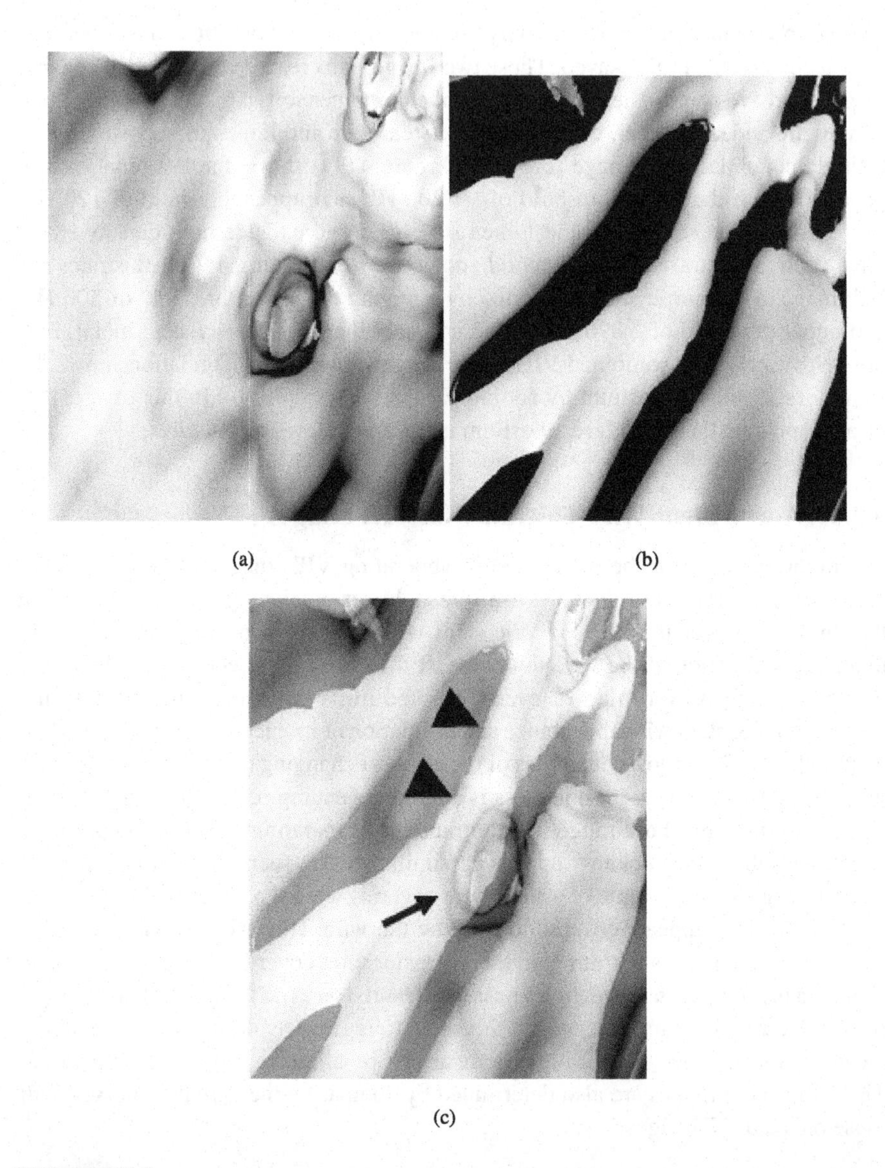

FIGURE 4.9
VIE view of aortic stent wires related to the ostium. VIE images of the renal ostium (a) and aortic stent (b) were first generated separately in the same anatomic location. Adding these two images allow the viewer to view the aortic stent wires crossing the renal ostium (c). Arrow refers to the renal ostium, while arrowhead points to the stent wire.

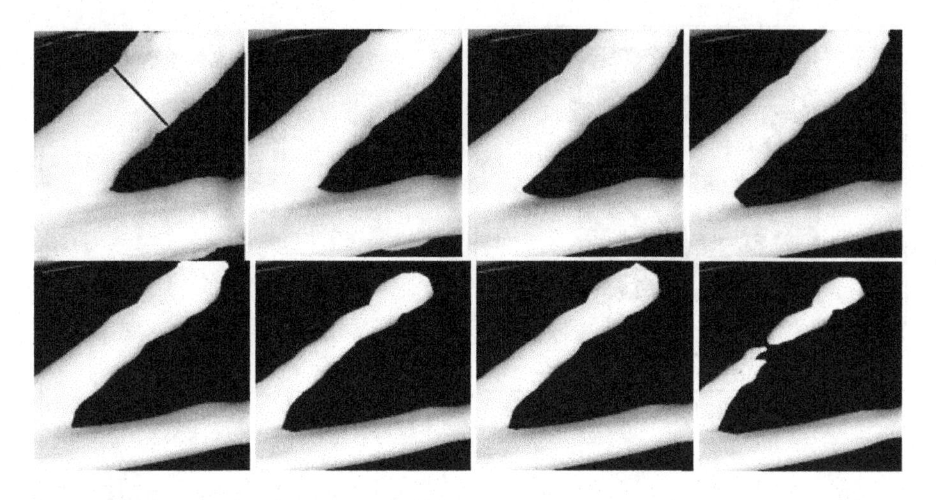

FIGURE 4.10
Effect of threshold selection on stent wire thickness and visualisation. A series of threshold ranges was applied to produce a set of VIE images of aortic stents with different thickness in diameter. The threshold ranges from 250 to 600 HU in steps of 50 HU. The corresponding diameters of stent wire ranged from 3.43 to 2.52 mm with a corresponding threshold range of 250 to 400 HU in the top row; 2.52 to 1.16 mm with a corresponding range of 450 to 600 HU in the bottom row of images. Note that the stent wire starts to become irregular or break when the threshold reaches 600 HU. The wire thickness measured on axial CT ranges from 1.64 to 1.96 mm. Therefore, the threshold range from 500 to 550 HU is the choice applied in this patient.

generates an endoscopic view at regular intervals. To cover the length of the aorta, a number of camera positions (10–15 virtual cameras) are used, depending on the anatomic regions to be covered, whilst up to 50 images are interpolated between each position. When played as cine viewing, the impression of a dynamic fly-through is obtained. With the latter step-by-step mode, each endoscopic view is set manually by the user to allow the visualisation of the particular anatomic structure of the abdominal aorta and its branches.[90]

On the initial viewing of intraluminal endoscopic images, there is little anatomy to orient oneself which initially makes the data difficult to interpret. The interpretation of VIE images should be viewed together with multiplanar reformatted images. A beginner may find it hard to explain what these endoscopic images represent, for example, a renal ostium is displayed as a hole in the aortic wall (Fig. 4.11) — is it a normal structure or an artefact? The simultaneous display of endoscopic and orthogonal images helps to confirm the anatomical location. Thus, for efficient analysis, the

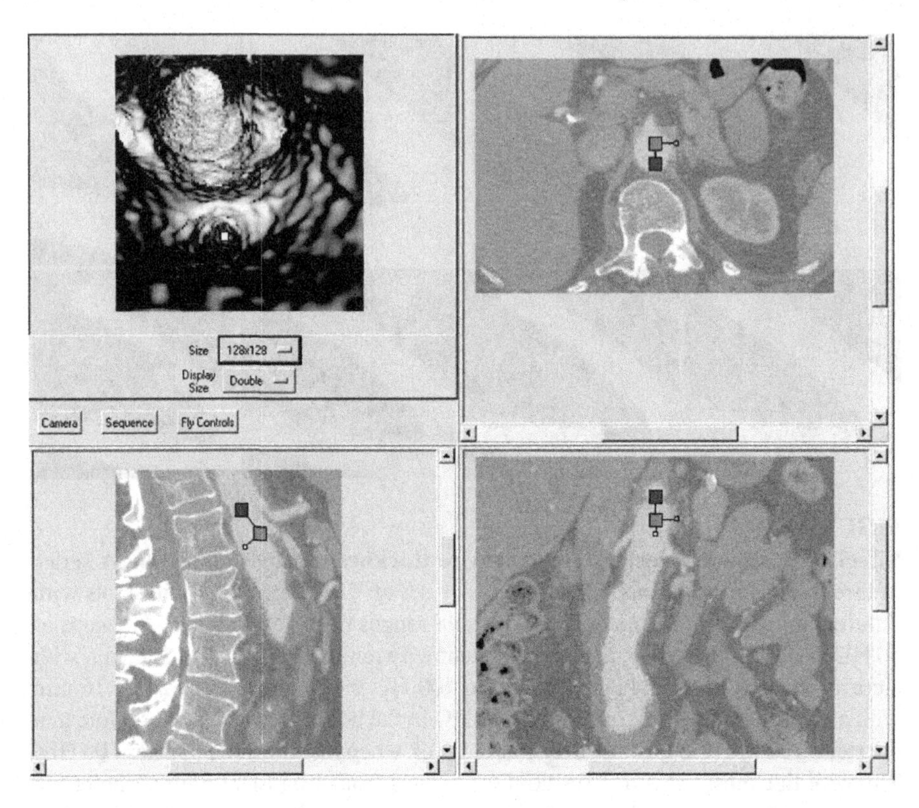

FIGURE 4.11
Interactive display of VIE images. The viewer's eye is looking inside the abdominal aorta towards the superior mesenteric artery (box in the top left image) with the top right image showing the VIE visualisation of the aortic ostium and intraluminal artery appearance. Axial, coronal and sagittal images help to adjust the viewing position and confirm the anatomical structure. Top row images: upper box indicates the virtual camera, while the lower box refers to the virtual eye. Bottom row images: upper box indicates the virtual eye position, while the lower box refers to the virtual camera.

computer screen simultaneously displays the endoluminal view with the corresponding orthogonal reformats to allow for the orientation, navigation and correlation of findings. With experience, familiarity with VIE images is increased and the processing speed enhanced and time shortened. With current fast-speed workstations, it may take about 10–15 min to produce VIE images from the coeliac axis to the common iliac arteries.

4.3 Optimal CT Scanning Protocols for VIE Visualisation

VIE has been reported to provide additional information when compared to conventional 2D and 3D visualisations in earlier studies,[90–93] and it was considered to be a valuable visualisation tool that assisted vascular surgeons to accurately evaluate the treatment outcomes of endovascular repair. Specifically, VIE was found to play a role in the following two aspects: providing unique information about the configuration and number of suprarenal stent wires crossing the renal artery ostia and assessing the morphological change of the renal artery ostia following suprarenal stent graft placement.[91,92]

As CT has become a routine technique in clinical practice, its application in endovascular repair of aortic aneurysm is being increased due to improved spatial and temporal resolution. Various examination parameters have been suggested by different groups for single slice and multi-slice CT VIE in aortic stent grafting,[4,92–96] as an optimal protocol that allows for the acquistion of acceptable diagnostic images with a minimal radiation dose to the patients.

In single-slice CT, with a given volume coverage rate, narrow collimation, high pitch helical scans provide better longitudinal resolution than wide collimation, low pitch scans.[4,94–96] For multislice CTA, similar results have been reported. Ideally, the minimum section thickness that allows for the visualisation of the aortic ostium as well as the aortic stent wire with fewer artefacts should be preferred. The most apparent of the artefacts is the stair-step artefact associated with surfaces or object borders inclined relative to the table translation direction. Stair-step artefacts characteristically deteriorate the appearance of 2D reformation and 3D render objects and may affect the accuracy of volume or diameter measurements of structures within the scanned volume.

With four-slice CT, for the purpose of visualising the aortic branch origin and stent wires with acceptable image quality observed on VIE, a scanning protocol of section thickness 2 mm, pitch 1.0 and reconstruction interval of 1 mm was recommended based on one of our early experiments.[95] With 64-slice CT, isotropic volume data can be acquired due to improved spatial and temporation resolution, thus a thinner slice thickness is most commonly used in clinical practice. This has been confirmed by our previous study showing that that a scanning protocol of section thickness 1.0 mm, pitch 1.4 and reconstruction interval of 0.5 mm is recommended as the optimal 64-slice CT VIE scanning protocol in post-aortic stent grafting.[4] Figure 4.12 demonstrates that the VIE image quality is independent of pitch values, while Fig. 4.13 shows the relationship between VIE image quality and slice thickness.

With the latest CT model, such as a dual-source CT scanner, a similar scanning protocol of slice thickness 1.5 mm, pitch 1.5 with 0.75 mm reconstruction was

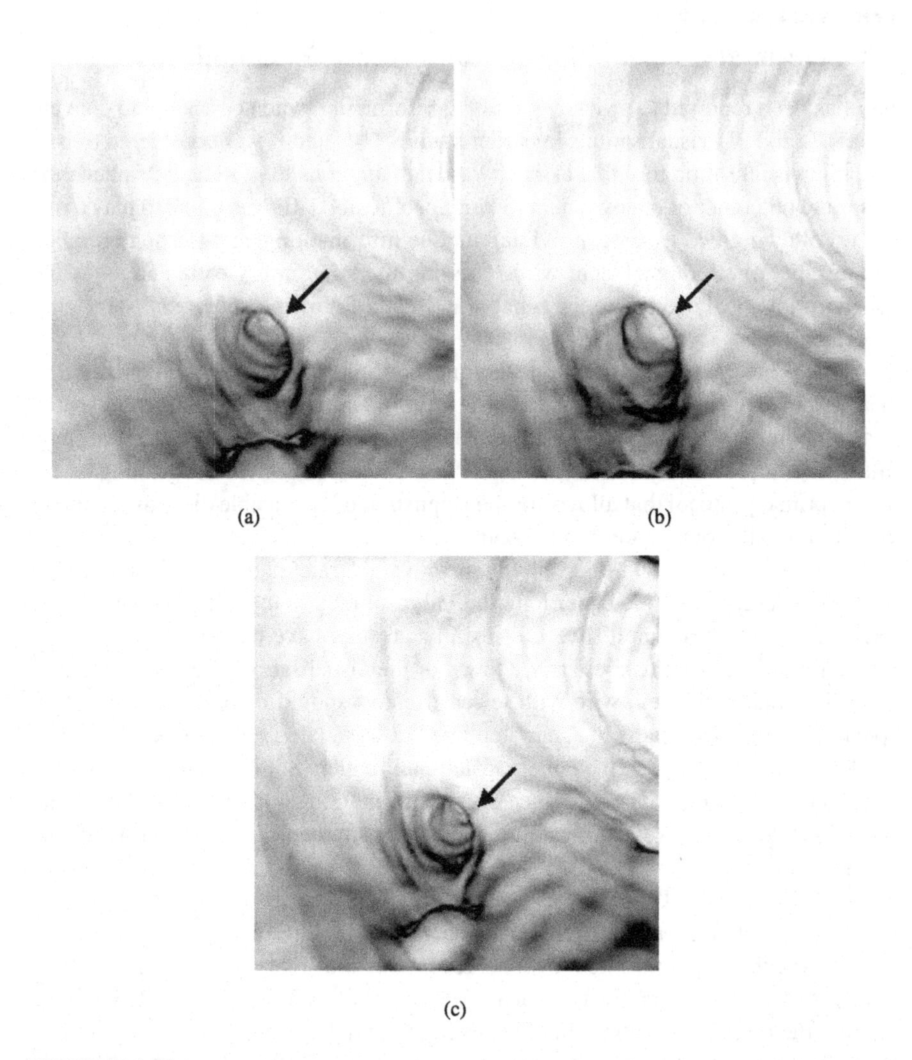

FIGURE 4.12
Relationship between pitch and VIE visualisation of renal ostium. VIE images of the right renal ostium were acquired with a section thickness of 1.3 mm, pitch values of 0.875, 1.25 and 1.75 and reconstruction interval of 0.6 mm ((a) to (c)). The renal ostium remains the same configuration on these three images and image quality is independent of pitch values. Arrows show the right renal ostium as observed on VIE images.

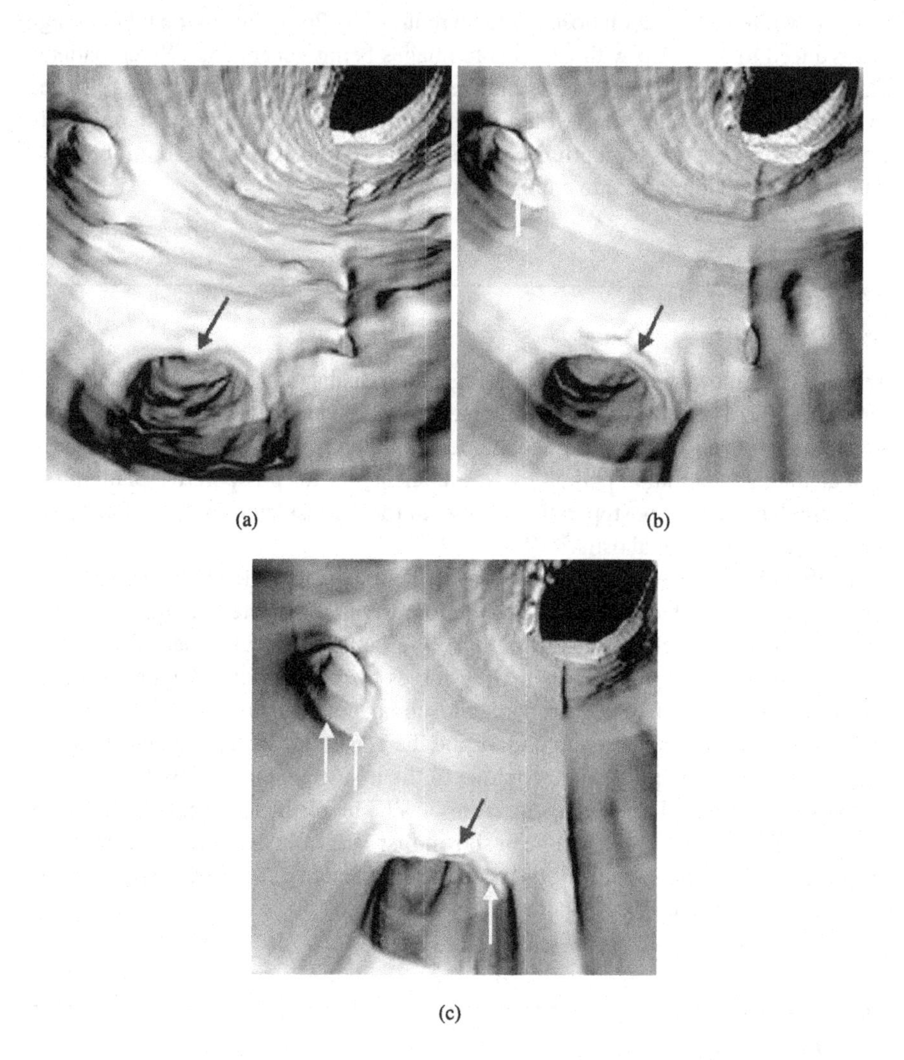

FIGURE 4.13
Relationship between slice thickness and VIE visualisation of renal ostium. VIE images of the superior mesenteric and renal ostia were acquired with a section thicknesses of 1.3 mm, 3.2 mm and 6.5 mm, pitch 1.25 and 50% of overlap ((a) to (c)). VIE image quality depends on the section thickness, as superior mesenteric and renal ostia become distorted when the section thickness reached 6.5 mm (c). Black arrows point to the superior mesenteric ostium, while the white arrows refer to the stair-step artefacts present in renal and superior mesenteric ostia.

reported, while the radiation dose could be reduced by 26.5% when the tube voltage was reduced to 100 kVp with acceptable images being generated.[97] Since patients with aortic aneurysms treated with endovascular stent grafts are normally followed up with a series of CT scans, which include three months, six months and yearly thereafter, it is necessary to optimise CT scanning protocols so that radiation risk to patients could be minimised to a greater extent.

4.4 Summary

Helical CTA allows real-time processing and visualisation of the 3D data. CT-generated VIE is a new postprocessing method which can be used as an adjunctive method to conventional imaging techniques in patients with aortic vascular disease such as the most common pathology, aortic aneurysm for preoperative planning of endovascular repair or postoperative follow up for the display of 3D relationship of aortic stents and the renal ostia.[88–93]

In this chapter, we have presented a methodology to generate VIE images of aortic aneurysms and aortic stents. In addition, we also indicated the importance of optimising CT protocols for the generation of VIE images of aortic aneurysms, aortic ostia and intraluminal stents. This mainly serves the purpose of reducing the radiation dose to patients while acquiring diagnostic images. The method developed here will help clinical investigators to understand the generation of VIE images in aortic stent grafting and to apply this technique to the assessment of endovascular repair of aortic aneurysms. The diagnostic value of VIE in aortic stent grafting has been investigated clinically, and it has been shown that VIE offers additional information to conventional visualisations for the accurate assessment of the treatment outcomes of endovascular repair.[86, 89–93, 98]

4.5 Questions

1) What is VE?
2) What are the different ways to obtain data to generate VIE images?
3) Name two aspects in which VIE has become more useful for vascular surgeons.
4) Why is it important to select an optimal threshold in VIE?
5) Give one reason to explain the increased application of CT in the endovascular repair of aortic aneurysms.

5

Treatment of Aneurysms

CONTENTS

5.1 Introduction 55
5.2 Open Surgery 56
5.3 Minimally Invasive Techniques 56
5.4 Medical Image Visualisation 57
5.5 Technical Limitations 60
5.6 Medical Imaging and Geometrical Reconstruction 60
5.7 Conformance with Preliminary Concept 64
5.8 Summary 66
5.9 Questions 67

5.1 Introduction

Most people with AAAs do not have aneurysm-related symptoms and the diagnosis therefore mainly depends on incidental findings. As most AAAs are asymptomatic, it is difficult to estimate their prevalence, but screening studies in the UK reported an estimation of a prevalence of 1.3–12.7% depending on the age group studied.[99] The incidence of symptomatic AAAs in men is approximately 25 per 100,000 at age 50, increasing to 78 per 100,000 in those older than 70 years.[99] The implementation of a national screening program for AAAs is recommended with aim of reducing the mortality.

Once an aneurysm has been detected by routine physical examination and radiographic studies, the risk of rupture is weighed against the risk of surgical repair for each patient. The major determinant for the risk of rupture is the aneurysmal diameter. In the absence of natural history data concerning patients with aortic aneurysms, the risk of rupture is estimated from the respective diameters of the abdominal aorta. The risk of operative complications is determined not only by age, cardiac and pulmonary function, but also by the extent of the aorta involved.

Several studies indicate that without surgery, the five-year survival rate for patients with aneurysms larger than 5 cm is about 20%.[99] Elective surgery is recommended in current clinical practice for patients with aneurysms larger than 5.5 cm in diameter and with aneurysms larger than 4.5 cm in diameter that have increased by more than 0.5 cm in the past six months.[100] Current guidelines from the Vascular Society and the National Screening Committee recommend that patients with asymptomatic aneurysms of less than 4.5 cm in diameter should be followed up with ultrasonography every six months, and aneurysms of 4.5–5.5 cm in diameter should be followed up every three to six months.[100]

5.2 Open Surgery

Definitive therapy for aortic aneurysms is to prevent aneurysm rupture, for example, by the placement of the dilated segment of the aorta with a prosthetic graft. To determine whether an individual patient is a candidate for graft replacement, many factors need to be considered, including the risk of aneurysm rupture, life expectancy, anticipated quality of life after the operation and the risk of surgical treatment.[101]

For more than half a century, open surgical repair has been used as the gold standard to treat AAAs with a high degree of success, and it is still widely performed in many clinical centres. The basic goal of surgical repair is the exclusion of the aortic aneurysm from the arterial pressure with the preservation of blood flow to the pelvis and legs via an implanted vascular conduit (usually a synthetic fabric or expanded polytetrafluoroethylene). This is usually achieved by an incision of the aneurysm sac, removal of mural thrombus, ligation of patent branch vessels (such as the superior mesenteric or inferior mesenteric artery) arising from the aneurysm sac, selection of a graft of appropriate size and shape, suturing anastomosis of the graft to the artery at proximal and distal segments to the aneurysm and finally, the closure of the decompressed aneurysm sac over the synthetic graft material.

This is a major and invasive operation by any criteria and the overall operative mortality for elective surgical repair is 4% or less, but can be as high as 8.4% depending on the experience of operating centres and patient's cardiovascular condition.[102,103] The major causes of peri-operative morbidity are cardiovascular, haemorrhagic and septic complications.[104] The risk of peri-operative death is more than 5% and may be as high as 65% in patients with co-morbid diseases such as coronary artery disease and renal failure.[105] Five-year survival after surgery is approximately 65–70%, with the major cause of long-term morbidity and mortality attributable to cardiac disease.[104,106] The long-term (i.e., 10-year) complications related to the surgical repair of aortic aneurysms are aneurysm formation at the surgical anastomoses (4%), recurrent true aneurysms (5%), graft occlusion (3%) and graft infection, aortoenteric fistula or both (combined incidence 5%).[107,108] In addition, recovery rates are in the region of several months, resulting in a reduction in quality of life for the patients for up to three months post-operatively.[109]

5.3 Minimally Invasive Techniques

In an attempt to reduce the surgical risk in patients with accompanied medical conditions, less invasive methods of repair have been considered. Minimally invasive techniques emerged about two decades ago for the repair of aneurysms and technical developments allowed them not only limited to the infrarenal aorta, but also extended to the suprarenal aorta.[110–112] Instead of graft replacement via an abdominal or flank incision, a thin-walled prosthesis is compressed into a catheter, introduced into the femoral artery via a limited groin incision and advanced into the aorta to exclude the aneurysm from the systemic circulation.

Endovascular repair of AAAs, as evaluated by the endovascular aneurysm repair (EVAR 1) and DREAM trials, has a lower 30-day mortality than conventional open repair.[113,114] The EVAR 1 trial recruited 1,082 patients, representing one of the largest planned trials of endovascular versus open surgery repair of AAAs.[113] It has been reported that there is a clear short-term benefit of EVAR, with 1.7% of patients dying within 30 days compared with 4.7% of those allocated open surgery. In addition, the 30-day and in-hospital mortality of EVAR was at least two-thirds lower than that of open repair. Similarly, the DREAM trial comparing open and endovascular repair of AAAs concluded that endovascular repair is preferable to open repair over the first 30 days after the procedure.[114] These randomised controlled trials indicated that in patients who are candidates for both open surgery and EVAR, endovascular repair leads to lower rates of operative mortality and complications and the significant reduction in the rate of systemic complications, thus it is a preferable approach in these patients.

The procedure of endovascular repair is performed less invasively when compared to the extensively invasive approach of open surgical repair. The prosthesis is anchored in place by means of balloon or self-expandable metallic stents. Perioperative imaging by means of 3D reconstructed CT scans and angiography are essential to guarantee the successful implantation of the prosthesis. The operation may be performed under regional or local anaesthesia. Theoretically, indications for endovascular grafting techniques could be extended to repair of aortoiliac, juxta-suprarenal and even thoracoabdominal aneurysms using this technique. Initially, only a limited number of patients could benefit from this method as a result of strict anatomic exclusion criteria. At both ends of the aneurysm, a "neck" of non-dilated vessel with a minimal length of 15 mm is essential for the secure fixation of the graft. Furthermore, aneurysms containing arterial branches that are indispensable for organ perfusion, such as juxtarenal and suprarenal used to be excluded from consideration for endovascular repair. Later, technical improvements in stent grafts have enabled it to be available to more patients, especially for those with short or complicated aneurysm necks.[115–117]

5.4 Medical Image Visualisation

Unlike conventional graft procedures, the success of endovascular stent graft repair cannot be ascertained by direct examination and thus relies on imaging assessment. While conventional angiography has been losing its dominant role for arterial imaging, spiral CT has been confirmed as the best single imaging technique for both preoperative patient assessment and aortic stent graft surveillance.[118,119]

Spiral CT has dramatically improved the performance of CT by converting a 2D modality into true 3D imaging, thus enabling the development of new applications involving volumetric imaging, such as CTA. In turn, CTA has been complemented by a parallel development of postprocessing methods to create a 3D representation of anatomical structures.[120,121] Among these reconstructions, VE is a unique visualisation tool which provides intraluminal views of the hollow organs, and is primarily used in the fields of colonoscopy and bronchoscopy.[122,123] A few studies have applied this technique to evaluate vascular diseases with VIE (Section 3.2.2).

Preliminary reports of VIE have shown the promising role of this technique in endovascular aneurysm repair.[120,121] Later studies with the use of VIE in the evaluation of patients treated with stent grafts have demonstrated that VIE provides advantages over traditional 2D visualisation methods.[124–127]

The primary clinical application of VIE in aortic stent grafting with the suprarenal component was found to be in the 3D visualisation of stent wire–renal ostia relationships, according to our previous experience[124–126] (Fig. 5.1). VIE is regarded as a valuable technique in this regard and superior to other imaging techniques because it offers a clear 3D intraluminal view of the stent wires and their position relative to the renal ostia. The long-term effect of suprarenal stent grafting is not well known, and we believe that the ability of VIE to characterise the stent wire–ostia relationship will prove a useful research and diagnostic tool that will enable the identification of any detrimental effect that suprarenal stent struts may have on the renal artery ostium and hence on renal function.

Another potential application of VIE in aortic stent grafting is to evaluate the treatment outcomes of fenestrated stent grafting, which represents technical developments over conventional endovascular repair with the aim of treating infrarenal aneurysms that have short aneurysm necks. The fenestration involves creating an opening in the graft fabric to accommodate the orifice of the vessel targeted for preservation. The initial animal feasibility study of fenestrated endovascular grafting was reported in 1999,[128] which led to the successful implantation in human subjects.[129,130] The fixation of the fenestration to the renal and other visceral arteries can be provided by the implantation of bare or covered stents across the graft–artery ostia interfaces so that a portion of the stents protrudes into the aortic lumen. Therefore, there are concerns about the loss of the target vessel resulting from the

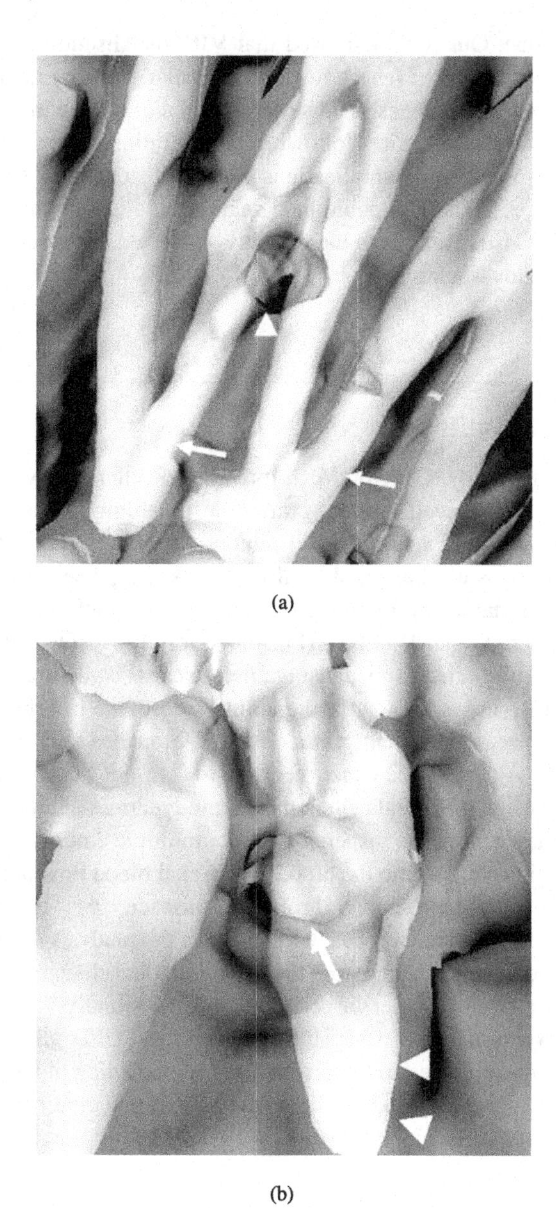

FIGURE 5.1
VIE of aortic ostia and stent wires. The right renal ostium (arrowhead in (a)) is crossed by two suprarenal stent wires (arrows), while the superior mesenteric artery (arrow in (b)) is encroached by a single stent wire (arrowheads).

fenestrated technique. Our results showed that VIE visualisation provides insights into the treatment outcomes of fenestrated endovascular grafts by demonstrating the intraluminal appearances of fenestrated stents in terms of unique information on the final configuration of the stent wires following a fenestration procedure (Fig. 5.2) and detecting any post-procedural complications such as distortion or deformity of the stent wires (Fig. 5.3). Thus, VIE could be a valuable and complementary technique to conventional 2D visualisations to identify any suspected abnormalities associated with fenestrated endovascular grafts.[127]

5.5 Technical Limitations

Conventional angiography, CTA or MRA provide excellent views of anatomical structures of the aorta as well as stent grafts, thus enabling the assessment of the diameter of aneurysms and stent grafts in relation to the aortic branches. CTA not only identifies artery wall changes, but also detects the presence of plaques and measures aneurysm diameters for surgical planning and follow-up of patients after endovascular stent graft repair[119, 124–127] (Fig. 5.4). Despite satisfactory results having been achieved, CT is limited to visualising anatomical or structural changes of the aorta or stent grafts, and is lacking in the ability to provide information about the haemodynamic impact of the stent grafts on aortic branches after the implantation of the stent grafts. Although the exact mechanisms are not known, it has been reported that the placement of stents alters the haemodynamics and this coupled with wall movement may lead to the dispersion of late multiple emboli.[131] The complex structures that are introduced into the blood flow (renal blood flow in the fenestrated repair) may enhance the biochemical thrombosis cascade,[132, 133] as well as directly affect the local haemodynamics. Thus, CFD enables the analysis of haemodynamic changes of the blood vessel, even before the morphological changes such as stenosis or occlusion to the renal or other visceral arteries are actually formed. Therefore, compared to conventional image visualisation methods, CFD allows for an early detection of abnormal changes and improves the understanding of the treatment outcomes of endovascular repair, so that the prevention of potential complications and better patient management can be achieved.

5.6 Medical Imaging and Geometrical Reconstruction

Computational methodology for cardiovascular systems begins with the construction of a computer model based on human imaging studies using contrast-enhanced CT or

(a)

(b)

FIGURE 5.2
VIE of fenestrated stents. A circular appearance of the left renal stent is noticed on VE image (a). Slightly distorted appearance (arrows in (b)) is observed on VE image at the inferior end of the right renal stent due to the balloon flaring effect after a VE operation. Arrowheads indicate artefacts caused by metallic stents.

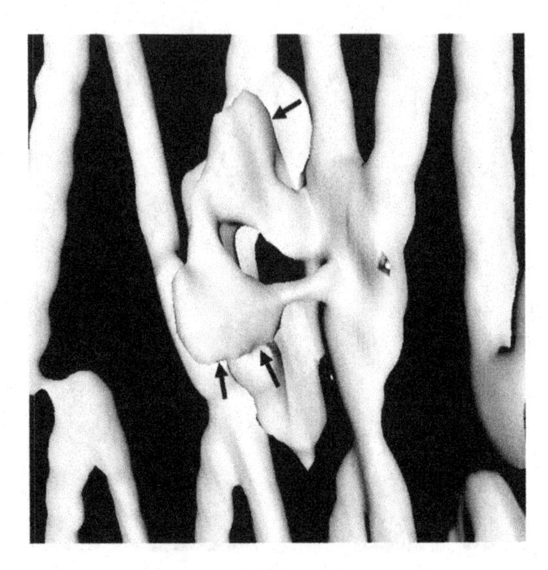

FIGURE 5.3
VIE of fenestrated stents with deformity. A deformed fenestrated left renal stent (arrows) is observed using VE.

MRI.[134, 135] The relevant vascular anatomy such as abdominal or thoracoabdominal aorta is imaged, and a computer model of the anatomy is constructed (Fig. 5.5). Using the anatomical 3D geometric model, FEA is performed to model interactions between the device and the circulatory system.[136]

Computer simulation of stents and stent grafts in anatomically and physiologically accurate patient models involves three steps: (1) the blood flow in the circulatory system, (2) the mechanical behaviour of the vessel wall and surrounding tissues under pulsatile and non-pulsatile loading and (3) the mechanical behaviour of the device. In order to avoid the mechanical failure of the devices and to guarantee their long-term efficacy and durability, the interactions of endovascular stent grafts with the arterial wall must be studied in detail. Model-based computer simulations can supply the necessary tools to improve stent grafts design and test the performance of endovascular stent grafts in a 3D patient-specific basis.

The CFD analysis of the flow patterns before and after endovascular repair and the calculation of the resulting shear stress enable the understanding of the stent graft performance. Early studies focused on the models of the interactions between the solid mechanics of the bloodstream lacked full volumetric analysis and were limited to idealised planar configurations.[137,138] Later reports showed the development of several important technologies including patient-specific cardiovascular modelling,[134] new methods for direct 3D geometric modelling[135] and 3D finite element

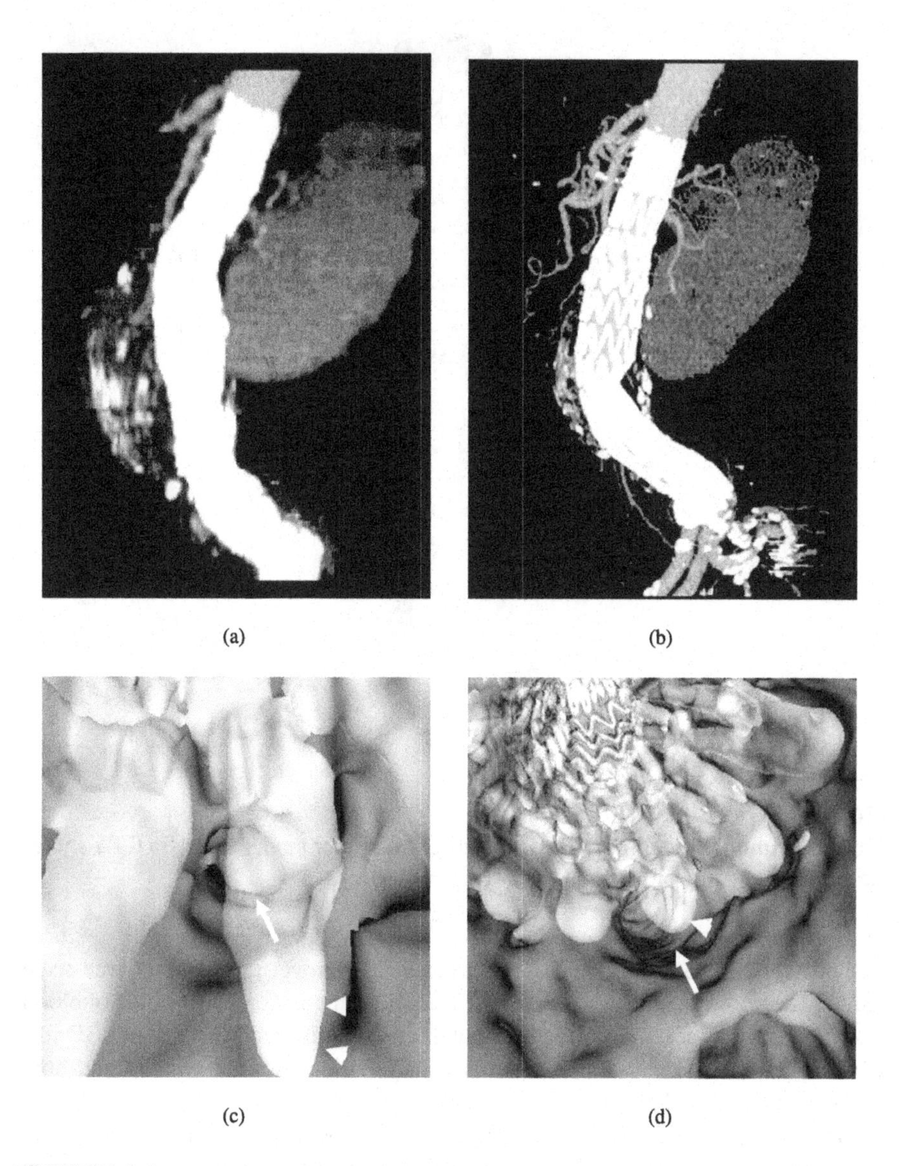

FIGURE 5.4
CTA follow-up of endovascular aneurysm repair. An aortic aneurysm was treated with a suprarenal stent graft (a) and evaluated at 36 months. Stent migration of 10.2 mm was noted in the most recent CT image (b). VE at one week (c) post-implantation showed a stent wire crossing the superior mesenteric artery ostium (arrow), but at 36 months (d), the stent wires (arrowheads) had shifted.

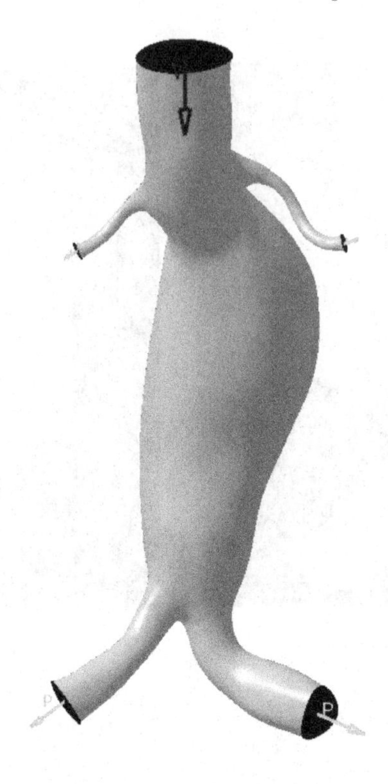

FIGURE 5.5
Computer modelling of aortic aneurysm. A computer model based on CTA shows the aortic aneurysm and normal artery branches.

methods for simulating blood flow,[139] as well as a novel method for large-scale FSI.[140] These improvements allow for the realistic computational modelling of blood flow in the human aorta. Thus, the computational solution produces realistic 3D images of the pulsatile blood flow patterns with variations in flow velocity and flow patterns under normal physiological conditions (Fig. 5.6).

Shear stress or wall shear stress exerted on the aneurysm sac can be calculated (Fig. 5.7). Displacement forces acting on endovascular stent grafts placed to treat the aneurysm can be calculated for both abdominal and thoracic stent grafts.

Given the availability of these computational methodologies, future endovascular devices can be tested in 3D computational models that accurately reflect the *in-vivo* flow conditions, thus long-term durability will be tested in simulated models prior to the implantation of the devices in patients. This allows for improved endovascular device designs with enhanced long-term safety and effectiveness of the devices.

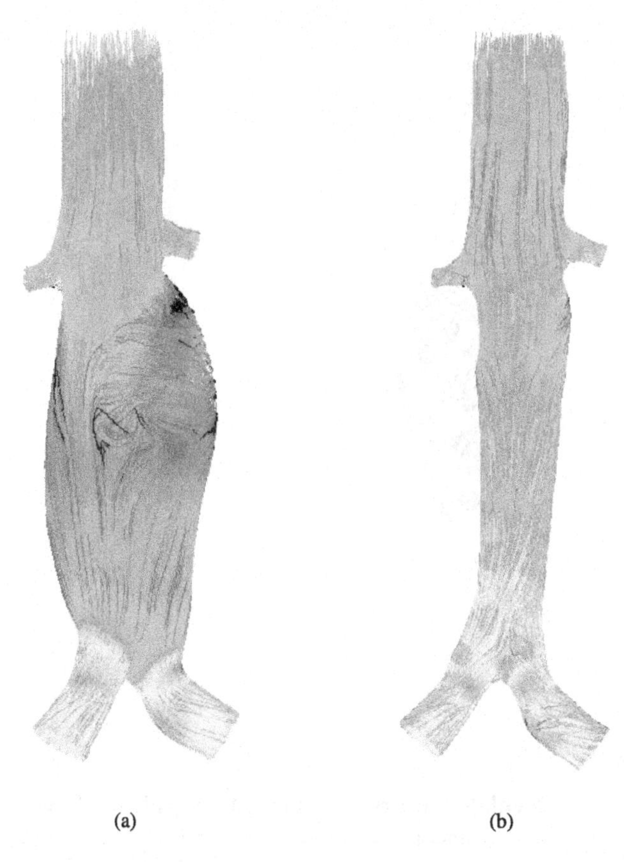

(a) (b)

FIGURE 5.6
Flow velocity measured at pre- and post-stent grafting. Turbulent flow is noticed inside the aneurysm before stent graft implantation (a), and flow becomes laminar following stent graft placement (b).

5.7 Conformance with Preliminary Concept

The presence of a non-biocompatible device inside an artery leads to inevitable inflammation and influences the fluid dynamic behaviour in the regions next to the arterial wall. Parts of the stent struts protruding into the lumen may induce the formation of vortices and stagnation zones which affect wall shear stress.[141] Stents grafts that are poorly matched to patient-specific vessels may pose vascular complications.[142, 143] In cases of serious conditions, subacute thrombosis due to inappro-

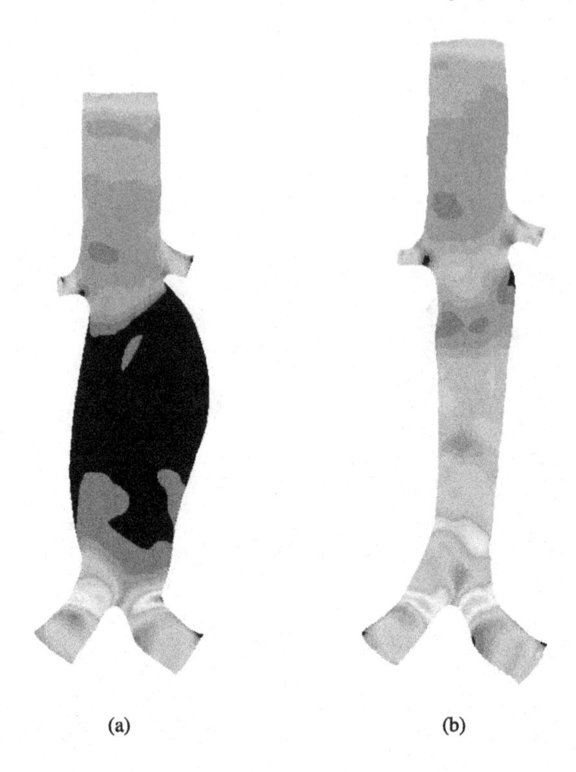

FIGURE 5.7
Wall shear stress calculated at pre- and post-stent grafting. Low shear stress is noticed at the location of aneurysm before stent grafting (a), but the shear stress increased significantly following stent graft placement (b).

priate use of stent grafts may result. Based on a patient-specific vascular circulatory system, simulated blood flow in an abnormal artery has been successfully performed to study complex fluid dynamics.[144, 145] CFD enhances our understanding of the interaction between stent grafts, flow analysis and the arterial wall; thus, better surgical intervention procedures can be developed, and optimal stent grafts can be achieved based on patient-specific conditions.

5.8 Summary

Endovascular aneurysm repair of AAAs has been confirmed as an effective alternative to open surgical repair due to its reduced invasiveness and lower mortality.

CTA is the preferred imaging modality for the pre-operative planning and post-operative follow-up of endovascular repair. Despite the high performance of CTA in aneurysm size measurements, the assessment of aneurysm extent and the detection of procedure-related complications such as endoleaks, CT lacks in providing information about the haemodynamic effect of stent grafts on aortic branches. This is complemented by computer modelling, which provides insights into the biomechanical behaviour of the stent grafts. A combination of medical image visualisation and computational simulation and modelling has the potential to improve stent graft design, thus successful treatment outcomes and better patient management can be achieved. Further studies are needed to develop a platform by combining 3D image visualisation with computer modelling to improve stent grafts design with the aim of achieving the long-term safety of endovascular repair of aortic aneurysms.

5.9 Questions

1) Describe one limitation in CT in image visualisation applications.
2) What is the advantage of using CFD over conventional image visualisation methods in endovascular treatments?
3) Explain how spiral CT has contributed towards the improved performance of CT.
4) List the three scopes of analysis that are involved in the computer stimulation of stents (and stent grafts) in anatomically and physiologically accurate patient models.
5) Explain why VIE is regarded as a valuable technique and superior to other imaging techniques for developing endovascular repair mechanisms.

6

Endovascular Stent Grafts

CONTENTS

6.1 Review of Device 69
6.1.1 What Is a Stent Graft? 69
6.1.2 Why Endovascular Repair? 69
6.2 Technical Developments 71
6.2.1 Suprarenal Stent Grafts 71
6.2.2 Fenestrated Stent Grafts 74
6.3 Technical Success 76
6.4 Long-term Outcomes 76
6.5 Computational Modelling 78
6.5.1 CFD of Suprarenal Stent Grafts 78
6.5.1.1 Configuration of Stent Wires Crossing the Renal Artery Ostium 78
6.5.1.2 Segmentation of CT Volume Data 79
6.5.1.3 Generation of Aorta Mesh Models 80
6.5.1.4 Simulation of Suprarenal Stent Wires Crossing the Renal Artery Ostium 81
6.5.1.5 Computational Two-Way Fluid Solid Dynamics 83
6.5.1.6 CFD Analysis 83
6.5.2 CFD of Fenestrated Stent Grafts 84
6.5.2.1 Simulation of Fenestrated Renal Stents 85
6.5.2.2 Numerical Verification 86
6.5.2.3 Computational Two-Way Fluid Solid Dynamics and Analysis 86
6.6 Summary 87
6.7 Questions 88

6.1 Review of Device

The use of stent grafts for treatment of AAAs in humans was first described by Parodi *et al.*[26] in 1991, who constructed devices from Palmaz stents (Cordis Endovascular, Miami, FL) and a standard woven polyethylene terephthalate surgical graft material. However, this series was limited to patients with AAAs that did not involve aortic bifurcation and iliac arteries. In 1994, Scott and Chutter[146] successfully placed a bifurcated stent graft designed by Chutter *et al.*[147] in six patients with AAAs. Parodi's work established the feasibility of the technique and stimulated worldwide interest. Endovascular stent grafts entered clinical trials in 1994.

6.1.1 What Is a Stent Graft?

A stent graft is an intraluminal device that consists of a supporting framework (made of metal such as stainless steel or nitinol) and a synthetic graft material (Fig. 6.1). Stent grafts can be either self-expanding or balloon-expanding, depending on the type of metal used in the stent. The stent may be located inside, outside or within the graft material, and it may be along the entire length of the graft or restricted to the ends. To deliver the stent graft through a small vascular access (most commonly via the femoral artery), the device is compacted into a catheter or compressed into a sheath. With the use of image guidance, the device is advanced into an appropriate location in the aorta from a remote access site and deployed.

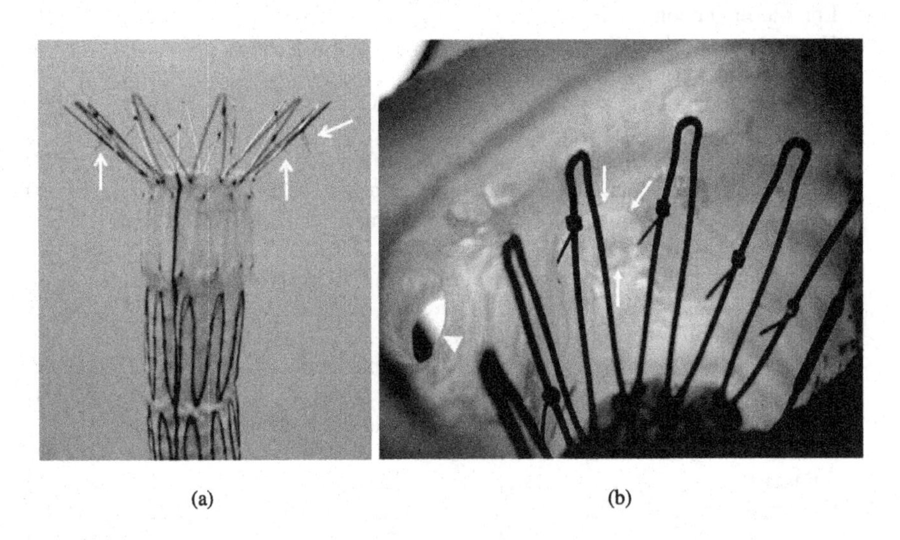

FIGURE 6.1
Example of a stent graft. A commercially available Zenith stent graft (by William A. Cook Australia Pty. Ltd.) consists of polyester graft with a sewed-on, self-expanding stainless steel wire frame. Beneath the main body of the stent graft, a separate "leg" is seen (a), which is inserted into the main body. Small hooks (arrows in (a)) welded into the wireframe make it possible to anchor the stent graft. Once deployed inside the aorta, the suprarenal component is placed above the renal artery to allow perfusion to the kidneys (arrows in (b)). Arrowhead indicates the superior mesenteric artery.

6.1.2 Why Endovascular Repair?

The ultimate goal in the treatment of an aortic aneurysm is to exclude the aneurysm from the aortic bloodstream without interfering with limb and organ perfusion. The risk of rupture of the aneurysms has to be weighed against the inherent risk of surgical repair. The larger the aneurysm, the greater the risk of rupture it may carry. Generally, the repair of AAAs with a diameter exceeding 5.0 cm is considered suitable. Cronenwett *et al.*[148] demonstrated that AAAs as small as 4.0 cm in diameter may rupture at a rate of 20% per year if hypertension is present. Thus, when the morbidity and mortality of surgical repair of aortic aneurysm can be further reduced, surgical intervention may be considered for the mere presence, and not the size, of an AAA.

The current technique of placing a prosthetic graft into the opened aneurysm and suturing it to "normal" aorta above and below requires extensive intra-abdominal or retro-peritoneal dissection, as well as the interference of blood flow during the completion of the anatomoses. Cardiac and respiratory impairment, renal failure, liver, intestinal and spinal cord ischaemia, ilieus, bleeding and coagulation disorders are all complications that can be attributed directly to the surgical procedure. The majority of these problems may be avoided when major surgical dissection and prolonged cross clamping are avoided by transfemoral intraluminal placement of the stent graft. The presence of the so-called hostile abdomen, with multiple peritoneal adhesions, prior vascular reconstructive procedures, malignancy or abdominal wall stomas, can greatly increase the difficulty of conventional aneurysm repair, but does not interfere with endovascular techniques. Furthermore, the incidence of post-operative impotence will be eliminated by avoidance of periaortic dissection, and the feared complication of aorto-enteric fistula should be prevented as the prosthetic material is never exposed to retro- or intraperitoneal structures. As is true for other minimally invasive techniques, hospitalisation time and convalescent period after discharge can be expected to be greatly reduced (from 7–10 days to 2–3 days), implying a decrease in overall cost as compared with open techniques.[149,150]

6.2 Technical Developments

The past decade has seen the development of a number of devices and techniques whose purpose has been a less invasive approach to the treatment of aortic aneurysms. These devices differ from one another with respect to the design features including modularity, the metallic composition and structure of the stent (Nitinol, stainless steel), the thickness, porosity and chemical composition of the graft material (synthetic/biological), the method of attaching the fabric to the stent (endo-stent configuration and exo-stent configuration) and the presence and absence of a penetrating method of fixing the device to the aortic wall with barbs and hooks.[151–154] However, the use of endovascular AAA repair remains limited in terms of the suitability of proximal and distal sealing and fixation. Estimates suggest that only 50% of patients with AAAs are candidates for endovascular repair with standard stent grafts on the basis of anatomic exclusion criteria.[155] A suboptimal neck makes stent graft fixation difficult and jeopardises the durability of a successful repair. The placement of an uncovered top stent over the renal artery ostia has been proposed to improve the fixation of stent grafts in patients with short and difficult aneurysm necks.[152–154]

Another technical development of endovascular repair is the recently introduced fenestrated stent grafts, which involves the creation of an opening in the graft material.[30,32,156] These technical developments make the application of endovascular stent graft repair available to more patients, especially for those with unsuitable aneurysm necks.

6.2.1 Suprarenal Stent Grafts

Suprarenal or transrenal fixation of aortic stent grafts, a modification of the commonly used infrarenal fixation, evolved to establish a more secure proximal fixation in patients with unfavorable proximal neck anatomy, thereby expanding the applicability of endovascular repair to a wider group of patients. Aneurysm neck

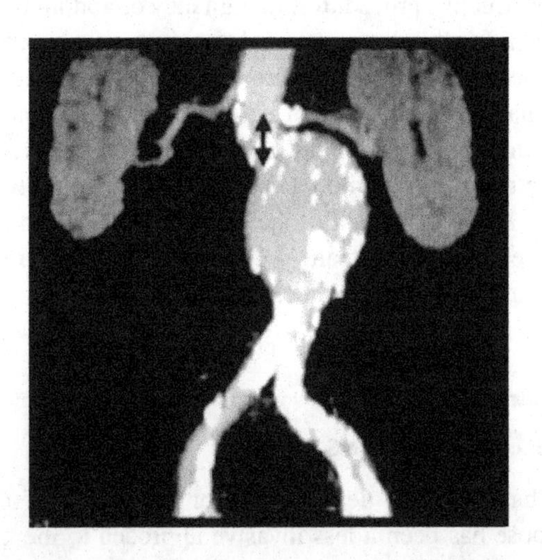

FIGURE 6.2
Short aneurysm neck. Three-dimensional coronal CT image shows an infrarenal aortic aneurysm with an aneurysm neck 9 mm in length (arrow). Extensive calcifications are observed in the aortic wall.

morphology is often the decisive factor for the feasibility of endovascular repair in infrarenal aortic aneurysms. Aneurysm neck length is particularly crucial because one third of patients subjected to endovascular repair have necks shorter than 15 to 20 mm.[157,158] A neck length shorter than 15 mm is usually considered insufficient for infrarenal endovascular fixation (Fig. 6.2). The migration of the proximal end of the stent graft has occurred in patients with short necks, but also may occur in patients with adequate neck morphology.[159]

Neck diameter is equally important. The mean diameter of the aneurysm neck in patients undergoing aneurysm repair is 24 mm.[160,161] Necks wider than approximately 28 mm are considered unsuitable for stent graft fixation because of the poor quality of the vessel wall (Fig. 6.3). Sever angulation (Fig. 6.4(a)), mural thrombus or atheromatous plaque (Fig. 6.4(b)) within the neck provides poor mechanical support for the stent and may increase the risk of an endoleak.[161,162] Calcification and exophytic plaques prevent the stent graft from anchoring tightly to the whole circumference of the neck and often cause proximal endoleaks.[161]

The combined factors mentioned above make current methods of treating infrarenal AAAs difficult. This means 30–40% of patients cannot be treated because of the risk of dislodgement of the stent or a proximal leak.[163] These complications often require the conversion to an open operation which carries a high mortality incidence of 50%.[164,165] Methods capable of dealing with these difficulties are needed, and a more proximal placement of the graft-anchoring stent may be a solution. The non-covered part of the stent graft can be positioned above the renal arteries to allow perfusion to kidneys, while the covered segment is placed below the renal arteries to avoid obstruction of renal perfusion (Fig. 6.5).

6.2.2 Fenestrated Stent Grafts

The principles of fenestration are to preserve blood flow to renal or visceral vessels and enhance stability by inserting stents into side branches to produce a durable relationship between the stent graft fenestration and the branch ostium (Fig. 6.6). It also requires the control and accuracy of deployment, especially rotational manoeuvrability to allow the exact positioning of the fenestration over the intended orifice. Fenestrated stent graft technology was developed in order to increase the morphological applicability of endovascular repair and offer an alternative to open surgery. The initial animal feasibility study of a fenestrated endovascular graft was reported in 1999,[30] which led to the successful implantation in human subjects.[30,32,156] Currently, the primary use of a fenestrated stent graft is to treat infrarenal aneurysms that have infrarenal necks <10 mm long.

The fenestration can be secured to the renal and other visceral arteries by the implantation of bare or covered stents across the graft–ostium interface so that a portion of the stent protrudes into the aortic lumen (Fig. 6.7).

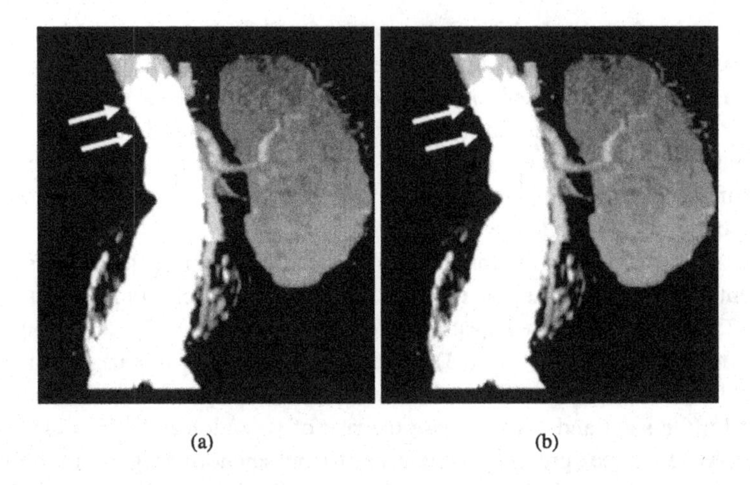

FIGURE 6.3
Wide aneurysm neck diameter. An infrarenal AAA with a neck length of 30 mm, but the neck diameter is 28 mm in width (arrow) (a). The patient underwent suprarenal endovascular stent placement (arrows in (b)) for the treatment of AAA due to the widened aneurysm neck (b).

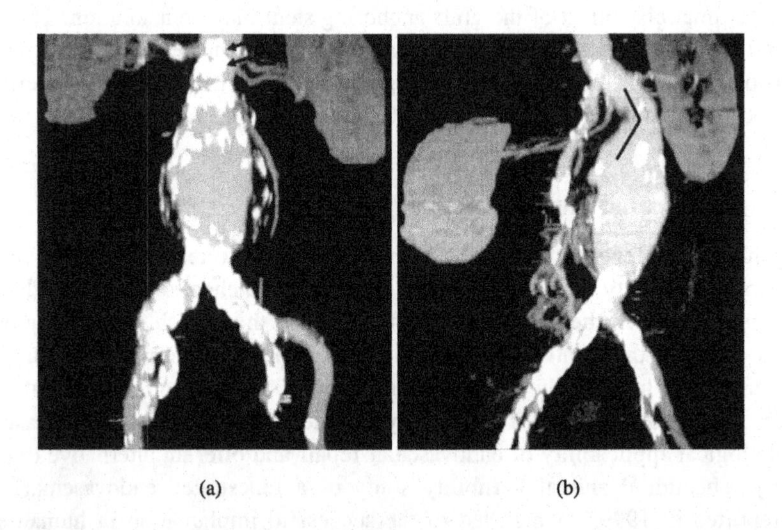

FIGURE 6.4
Complicated aneurysm necks. Coronal CT reconstructed images show extensive calcification present in the aneurysm neck (arrows in (a)) and severe angulation of the aneurysm neck (line measurement of angle in (b)) which prevent the patients from being candidates for conventional infrarenal endovascular repair.

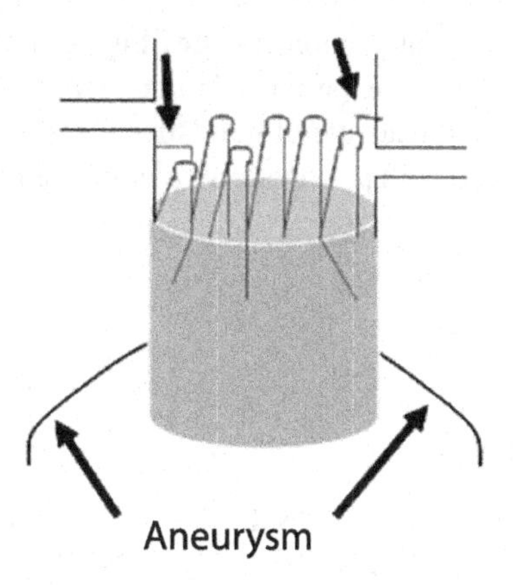

FIGURE 6.5
Principles of suprarenal fixation of stent grafts. An uncovered portion of the proximal stent protrudes over the renal arteries. Hooks and barbs engage the aortic wall (short arrows).

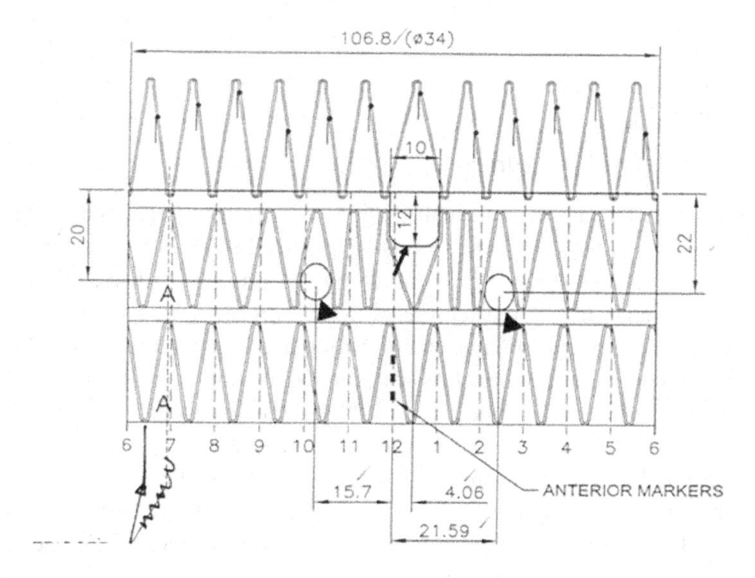

FIGURE 6.6
Planning diagrams for fenestrated repair of aortic aneurysm. An open view of upper portion of the stent graft shows showing a standard fenestration (arrow) and small fenestrations (arrowheads) that will be implanted in the superior mesenteric artery and renal arteries, respectively.

Accurate planning and construction requires high quality 3D CT imaging and good angiography. Since CT has experienced rapid technical developments over the last decade, especially with the advent of multislice CT[166] (Fig. 6.8), it has replaced conventional angiography in both pre-operative planning and post-procedure follow-up.

6.3 Technical Success

The 30-day mortality rate in large stent graft series ranges from 0.7% in low risk populations to 15.7% in high risk patients.[151,167] These statistics compare favourably with those associated with the surgical repair of AAAs.[168] Death during a stent graft procedure is rare. Randomised controlled trials reported that endovascular repair is preferable to open repair.[114,169] The 30-day mortality rate of endovascular repair is significantly lower than that observed in the open repair of patients who have an abdominal aortic aneurysm more than 5 cm in diameter. However, long-term follow-up is needed to determine the safety of this technique.

6.4 Long-term Outcomes

The physiologic effects of transrenal fixation on long-term renal function remain unknown. Investigators in several reports have described short to mid-term results for transrenal repair of AAAs.[170–172] These reports provide evidence which shows the safety of placing uncovered suprarenal stents over the renal arteries as assessed by the biochemical examination of renal function and imaging follow-up. However, there are still concerns about the safety of transrenal fixation as the long-term effect of this technique is still not fully understood.[173–175] The consequences of the suprarenal stent coverage of the renal arteries can manifest in various ways, e.g., interference with renal blood flow or renal function, decreased cross-sectional area of the renal ostium or a biological response of the aorta to the aortic stents. It has been shown that the renal artery ostium demonstrated morphological changes to variable degrees following suprarenal fixation; however, most of the changes did not represent a significant difference, and no renal dysfunction was observed. In patients with calcification in the renal artery ostium, a significant reduction of ostial diameters was seen in the right renal artery, indicating a potential atherosclerotic effect.[176]

Short- to mid-term outcomes of fenestrated stent grafting are satisfactory;[177,178] however, there are concerns about the patency of fenestrated vessels and interference

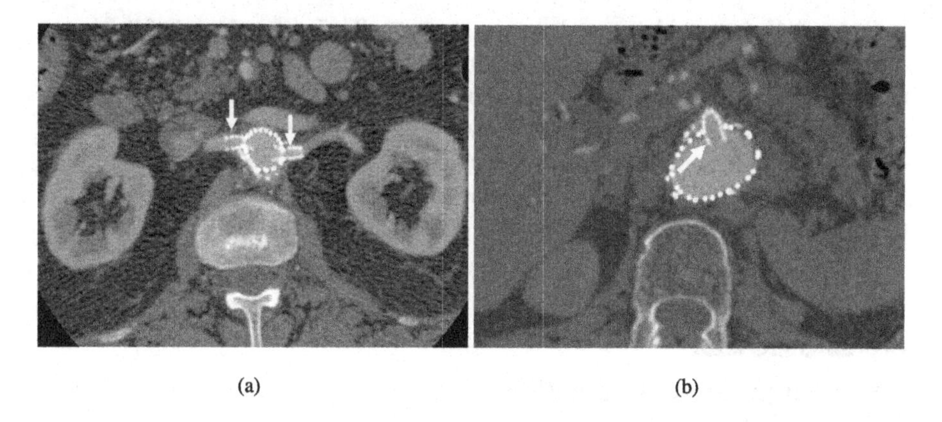

FIGURE 6.7
Fenestrated vessel stents. Two-dimensional axial CT images show that the small fenestrated stents (arrows in (a)) were inserted into the bilateral renal arteries and large fenestrated stent (arrow in (b)) in the superior mesenteric artery.

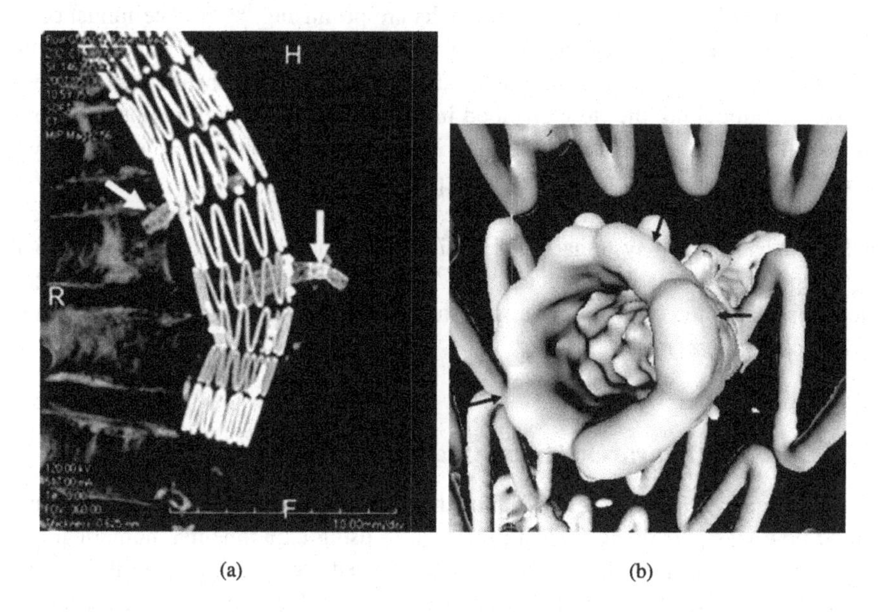

FIGURE 6.8
Fenestrated stent graft. A fenestrated stent graft was used to treat a patient with a short aneurysm neck. Renal stents were inserted into bilateral renal arteries (arrows in (a)) to provide support and allow renal perfusion. Intraluminal protrusion of the renal stent is clearly visualised on a VIE image (arrows in (b)).

of fenestrated stents with haemodynamics, as normally about one-third of the fenestrated stents protrudes into the aorta after implantation (Fig. 6.7).[179–181] Although the exact mechanisms are not known, it has been reported that the placement of stents alters the haemodynamics and this coupled with wall movement may lead to the dispersion of late multiple emboli.[33] The complex structures that are introduced into the blood flow (renal blood flow in the fenestrated repair) may enhance biochemical thrombosis cascades,[34,182] as well as directly affect the local haemodynamics.

6.5 Computational Modelling

The design and use of an endovascular stent graft requires an understanding of the expected forces which the stent graft will encounter once it is placed within an AAA. The deployment of a complex multicomponent endovascular device in the abdominal aorta is likely to alter the local haemodynamics and adversely affects the long-term performance of the device. Studies have been performed to investigate the effect of stent grafts in flow structures, and results are promising.[183–186] The influence of specific stent graft geometry on the flow pattern has been addressed using CFD with regard to the overall velocity profiles, wall pressure and wall shear stress. This involves the modelling the flow of blood in the stent graft by solving the equations of fluid flow via numerical methods. The increasing processing speed of computers makes CFD a relatively practical and inexpensive option.

The haemodynamics and biomechanics of an AAA following infrarenal endovascular repair has been systematically studied by researchers based on experimental or computational modelling studies.[187–190] The effect of suprarenal and fenestrated stents on renal blood flow has not been studied in depth, and this will be addressed in the following sections.

6.5.1 CFD of Suprarenal Stent Grafts

The representative study of investigating the effect of suprarenal stent wires on renal blood flow was performed by Liffman *et al.*[191] using experimental, numerical and analytical modelling methods. Their results showed the stent wire had little effect on renal blood flow through medium-sized vessels, and less than 1% reduction was observed as long as there is no build up of material on the wire. Since suprarenal stent wires cross the renal artery ostium in different configurations, depending on the individual patient situation, haemodynamic analysis based on realistic aorta models generated from the patient's data will reflect the real clinical situation. Consequently, results can be translated to clinical practice.

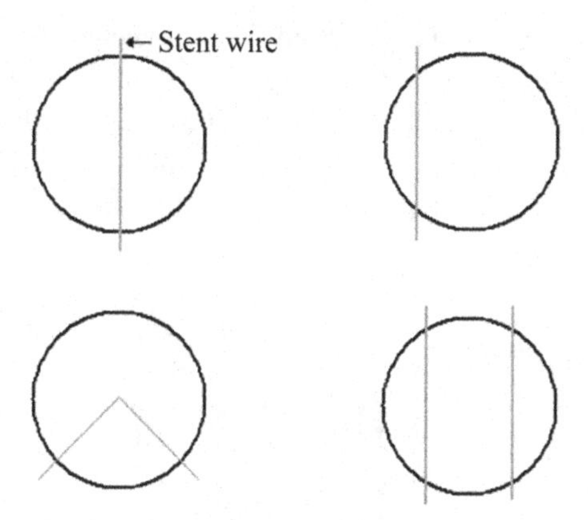

FIGURE 6.9
Configuration of suprarenal stent wires crossing the renal artery ostium. Top row: single wire crossing centrally and peripherally; bottom row: V-shape wires crossing centrally and multiple wires crossing peripherally.

6.5.1.1 Configuration of Stent Wires Crossing the Renal Artery Ostium

The generation of intraluminal images of the stent wires in relation to the renal artery ostium is performed using a CT number thresholding technique, which has been described.[51] The suprarenal stent wires have been reported to cross the renal artery ostium in four different configurations,[51,176,192] namely single wire crossing centrally, single wire crossing peripherally, V-shaped wire centrally crossing and multiple wires crossing peripherally. Figure 6.9 presents the diagrams of these four types of configurations.

6.5.1.2 Segmentation of CT Volume Data

Patients with AAAs undergoing endovascular repair are routinely scanned with CT imaging to determine the extent of aneurysm for surgical planning and follow-up of the treatment outcomes (monitoring aneurysm changes, detection of complications such as stent graft migration or endoleak). In order to produce realistic aorta models for haemodynamic analysis, the CT volume data must be postprocessed to extract the anatomic details for further analysis. The segmentation of CT volume data is performed with a semi-automatic segmentation technique involving CT number thresholding, region growing and objects creation and separation. For the generation of a 3D aorta model with the inclusion of only the main abdominal aorta and its branches, the lowest and highest CT thresholds are normally defined as 200 and 400 HU, respectively to remove all of the soft tissues, bone structures and stent wires, while keeping the contrast-enhanced artery branches; for the generation of a 3D aorta model with the inclusion of endovascular stent grafts, the lowest threshold is set at 500 HU to remove all of the soft tissues and contrast-enhanced vessels, while keeping the high-density stent wires. Figure 6.10 shows the 3D aorta models from the CT data of a selected patient pre- and post-suprarenal stent grafting.

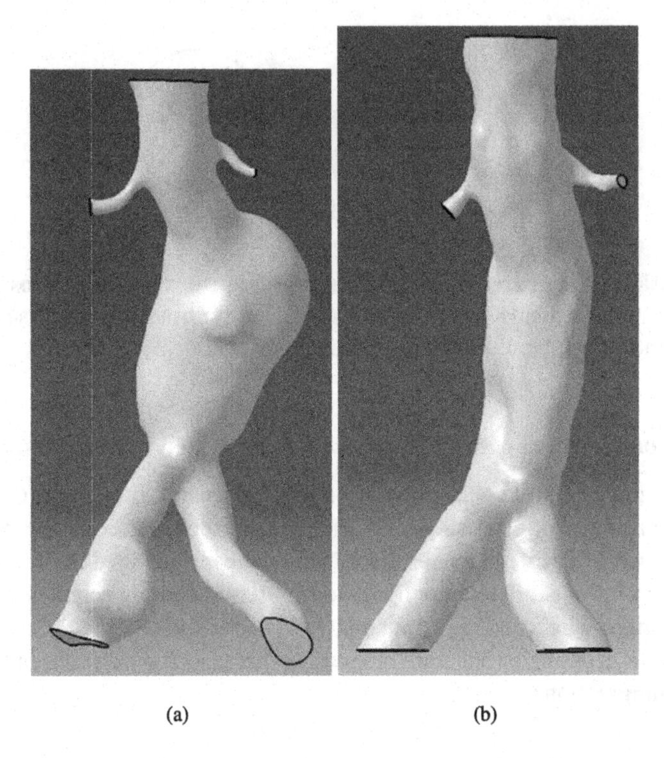

FIGURE 6.10
Realistic aorta models pre- (a) and post-stent graft placement (b) based on a sample CT data.

6.5.1.3 Generation of Aorta Mesh Models

After segmentation, 3D surface objects are created with the aid of computer software, and the 3D surface objects are saved in STL format, a common standard for computed-aided design and rapid prototyping. The "STL" file is then converted into CAD model files for the generation of aorta mesh models.

The aorta mesh model consists of three parts in each patient: part 1 refers to the blood flow model of pre- and post-stent grating, while part 2 indicates the artery

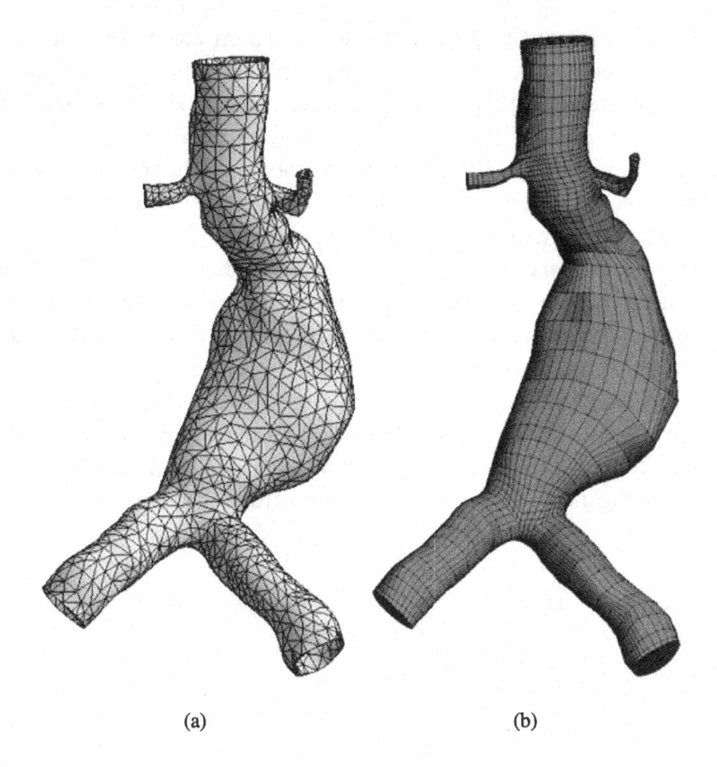

FIGURE 6.11
Aorta wall (a) and blood flow mesh (b) models.

wall model of pre- and post-stent grafting, and part 3 is the blood flow model of post stent grafting with the placement of the simulated suprarenal stent wires. For advanced mesh modelling, the blood flow model (part 1) is generated by hexahedral volume meshes, while the blood wall model (part 2) is generated by tetrahedral volume meshes. The blood flow model with the insertion of the suprarenal stent graft (part 3) is generated by tetrahedral volume meshes. The structure and fluid mesh models of a sample patient are shown in Fig. 6.11. The maximum elements of the blood wall model and flow model are composed of 17,247 and 82,650 elements, respectively.[52]

6.5.1.4 Simulation of Suprarenal Stent Wires Crossing the Renal Artery Ostium

Although the segmented AAA models are generated with CT number thresholding which focuses on the high-density stent wires, the detailed configuration of suprarenal stent wires crossing the renal artery ostium could not be displayed in the final mesh models. In order to simulate the intraluminal configuration of stent wires crossing the renal artery ostium, a few models with a simulated wire thickness of 0.4, 1.0 and 2.0 mm are generated respectively to reflect the actual situation of endovascular repair. Figure 6.12 shows meshing models of the suprarenal stent struts. The stent strut model is defined by tetrahedral volume mesh and mesh elements of stent strut model are between 9,790 and 23,471 elements.[52]

The stent strut model reflects the realistic clinical situation after the implantation of suprarenal stent grafts in patients with AAAs. The actual wire thickness of the suprarenal stent component is 0.4 mm, thus the simulated wire thickness is chosen to be 0.4 mm in diameter. As there is a potential opportunity for blood materials to build up on the stent surface over a certain period of time leading to the thickening of stent wires,[191] wire thickness of 1.0 and 2.0 mm are simulated, respectively, to demonstrate different wire thicknesses as shown in Fig. 6.12. Before simulating the stent struts crossing the renal ostium, the non-strut crossing models are first generated

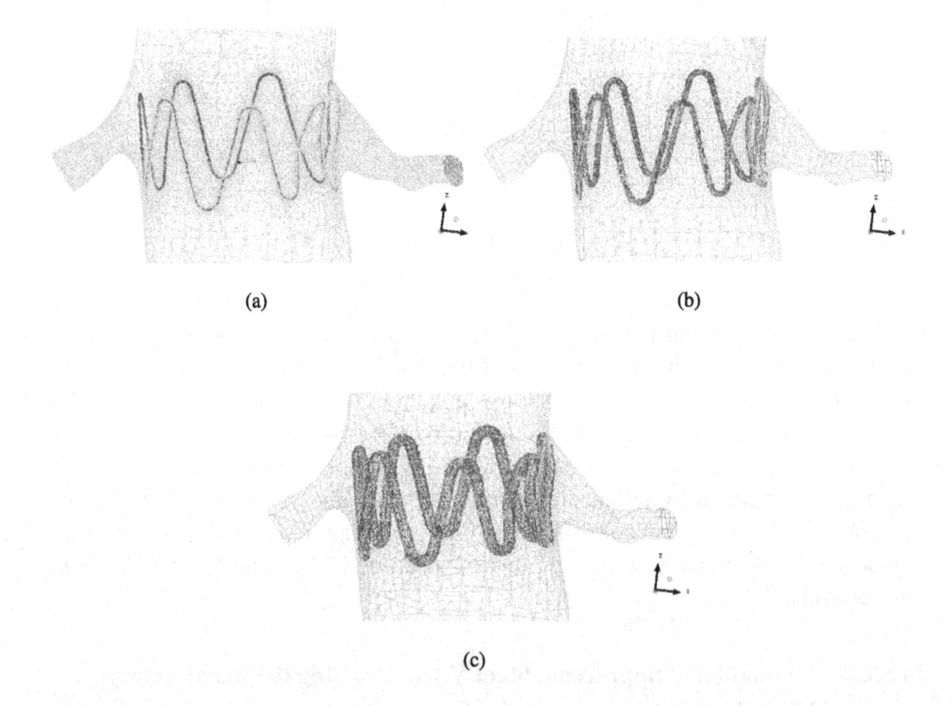

FIGURE 6.12
Mesh models of suprarenal stents. A stent wire thickness of 0.4 mm (a), 1.0 mm (b) and 2.0 mm (c) was simulated to cross the renal artery ostium in these models.

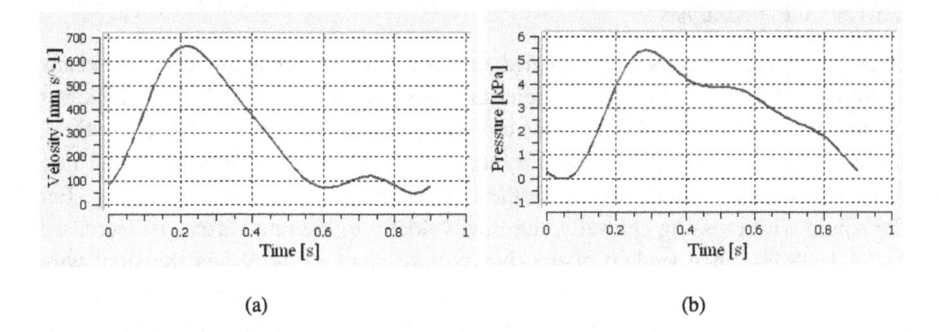

(a) (b)

FIGURE 6.13
Flow pulsatile (a) and time-dependent pressure (b) in different cardiac cycles.

as a reference for comparison with these models with the strut crossing the renal ostium at different configurations.

6.5.1.5 Computational Two-Way Fluid Solid Dynamics

To simulate the realistic situation in patients with AAAs treated with suprarenal stent grafts, the blood flow is started at the level of celiac axis, then inside the aortic aneurysm, and flows out to the renal arteries and common iliac arteries. The boundary conditions are time-dependent.[193] The velocity inlet (abdominal aorta at the level of coeliac axis) boundary conditions are taken from the referenced value showing measurement of the aortic blood velocity (Fig. 6.13(a)). A time-dependent pressure is also imposed at the outlets (Fig. 6.13(b)).

The fluid (blood) is assumed to behave as a Newtonian fluid, as this is known for the larger vessels of the human body. The fluid density is set to 1,060 kg/m^3 and a viscosity of 0.0027 Pa·s, corresponding to the standard values cited in the literature.[193] The flow is assumed to be incompressible and laminar. Given these assumptions, the fluid dynamics of the system is fully governed by the Navier–Stokes equations, which are shown as follows:

$$\text{Continuity: } \nabla \cdot \vec{v} = 0 \qquad \text{and} \tag{6.1}$$

$$\text{Momentum: } \rho\frac{\partial \vec{v}}{\partial t} + \rho\vec{v} \cdot \nabla\vec{v} = -\nabla p + \mu\nabla^2\vec{v} + \text{f in}^F\Omega(t)\,, \tag{6.2}$$

where $\vec{v}$ is the blood velocity vector, p is the blood pressure, ρ is the blood density, μ is the blood viscosity, f is the body force at time t acting on the fluid per unit mass, ∇ is the gradient operator and $^F\Omega(t)$ is the fluid domain at time t.

6.5.1.6 CFD Analysis

Haemodynamic analysis is performed in these realistic aorta models and results showed that the flow velocity to the renal artery is mainly determined by the thickness of stent wires and type of stent wires crossing in relation to the renal artery ostium. Flow velocity is slightly decreased by up to 5% with a wire thickness of 0.4 mm in all types of configurations except the type of single wire centrally crossing. For the single wire crossing centrally, the flow velocity to the renal artery is decreased by 21.1–28.9%, independent of the thickness of stent wires. When the stent wire thickness increases to 1.0 and 2.0 mm, flow velocity is decreased by more than 10% and as high as nearly 30% in most of the situations, indicating that the wire thickness is the determinant factor in the flow analysis[52] (Fig. 6.14).

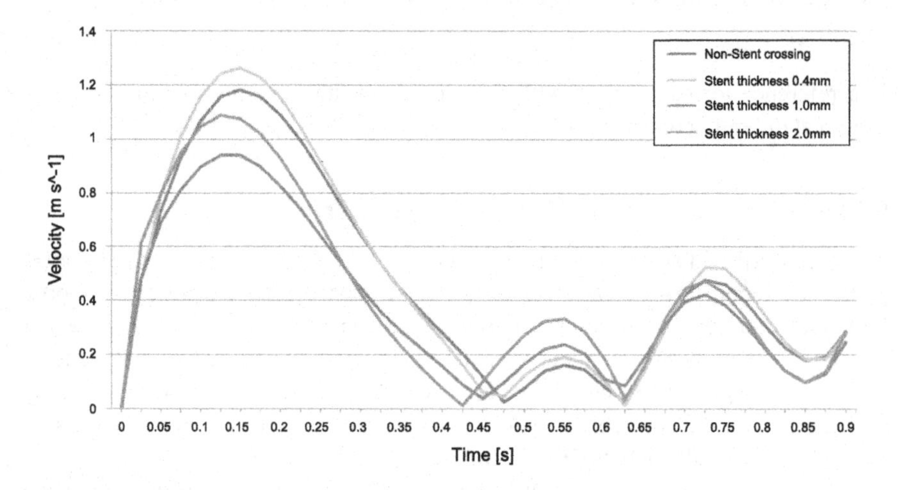

FIGURE 6.14
Haemodynamic effect of suprarenal stent wires on renal blood flow. Flow analysis in a patient without and with a single stent wire crossing the renal artery ostium peripherally. It is observed that when the stent wire thickness was increased to more than 1.0 mm, the flow velocity was decreased when compared to that observed without wire crossing and wire thickness of 0.4 mm.

6.5.2 CFD of Fenestrated Stent Grafts

Fenestrated stent grafts have been confirmed to be an effective alternative to conventional endovascular repair of AAAs (both infrarenal and suprarenal fixation), especially in patients with short or complicated aneurysm necks.[178] Early reports of fenestrated stent grafting are promising as the technique of endovascular repair has been further advanced and made available to more patients, especially for those with very short or hostile aneurysm necks.[177,178] However, there are concerns about the patency of fenestrated vessels and interference of fenestrated stents with haemodynamics, as a certain length of (approximately one-third, which ranges from 5–8 mm) of the fenestrated stents protrudes into the aorta after implantation.[180,181,194,195] Therefore, the investigation of the effect of fenestrated vessel stents on subsequent blood flow, particularly the renal artery flow or perfusion, is of paramount importance for improving the understanding of the long-term outcomes of the procedure.

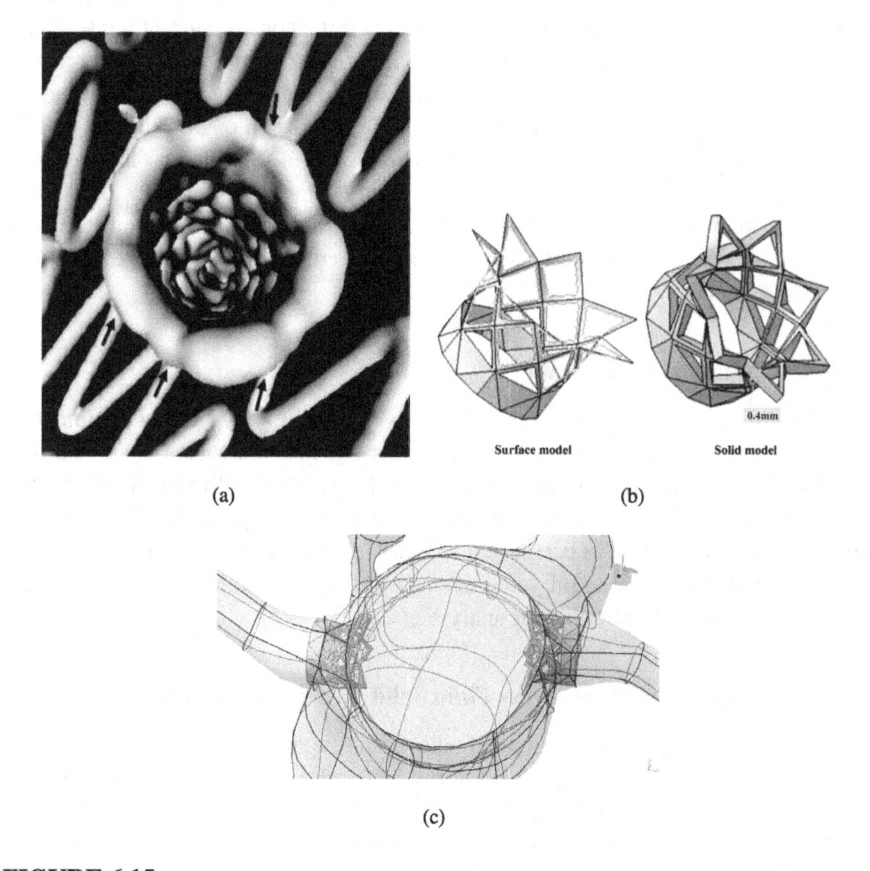

FIGURE 6.15
Simulation of intraluminal appearance of fenestrated renal stents. Appearance of an intra-aortic portion of the fenestrated renal stent (arrows in (a)) is displayed on a 3D VE image. The surface model of the fenestrated renal stent is simulated (b) and inserted inside the renal arteries with a protruding length of 5–7 mm into the abdominal aorta (c).

6.5.2.1 Simulation of Fenestrated Renal Stents

Although the segmented post-stent grafting AAA models are generated with CT number thresholding and other postprocessing methods focusing on the high-density stent wires, a detailed configuration of fenestrated renal stents inside the renal arteries could not be displayed in the final mesh models. To achieve this goal, fenestrated stent structures are simulated, which are later inserted into the aorta models to reflect the actual patient treatment. The models of the fenestrated stent wires are created by taking a reference from the intraluminal appearance of a fenestrated stent inside the renal artery visualised with 3D VE (Fig. 6.15(a)).[180,181] First, the renal artery diameter is measured and used it as the baseline for the construction of the scaffolding of the stent wires. Then, the structure profile of the stent wires is generated to produce the surface and solid models (Fig. 6.15(b)). Finally, the simulated model is inserted into the renal artery with an intro-aortic protrusion of 5–7 mm, as shown in Fig. 6.15(c). As the thickness of stent wires is about 0.4 mm in diameter, and the fenestrated renal stents consist of six to eight V-shaped metal wires protruding into the abdominal aorta with a length of less than 7 mm, according to our previous experience,[180,181] thus the simulated renal stents are created by reflecting the realistic patient's treatment.

6.5.2.2 Numerical Verification

In order to satisfy the criteria for mesh convergence, meshes for both fluid and solid domains have to be refined until mesh-density independence of the results is achieved. The maximum number of nodes per element is 18,020 and 71,921 for the artery wall mesh model and the blood flow mesh model, respectively.[196] A coupled fluid–structure simulation is performed at a variable time step with different cardiac cycles so that fluid forces and velocities across the fluid–solid interface could be demonstrated and calculated in our analysis.

6.5.2.3 Computational Two-Way Fluid Solid Dynamics and Analysis

Similar to what has been described above in the haemodynamic analysis of suprarenal stent wires, normal physiological haemodynamics should be considered for the 3D numerical simulations. This allows the study of the aneurysmal fluid mechanics by taking into account the instantaneous fluid forces acting on the wall and the effect of the wall motion on the fluid dynamic field. Fluid density and other characteristics are exactly the same as those described in the Section 6.5.1.5. The simulation of blood flow (fluid) is performed at different cardiac cycles.

Flow velocities to the renal arteries pre- and post-fenestration are calculated and compared in both pre- and post-fenestration, and our analysis shows that no significant interference with renal haemodynamic analysis is observed in the presence

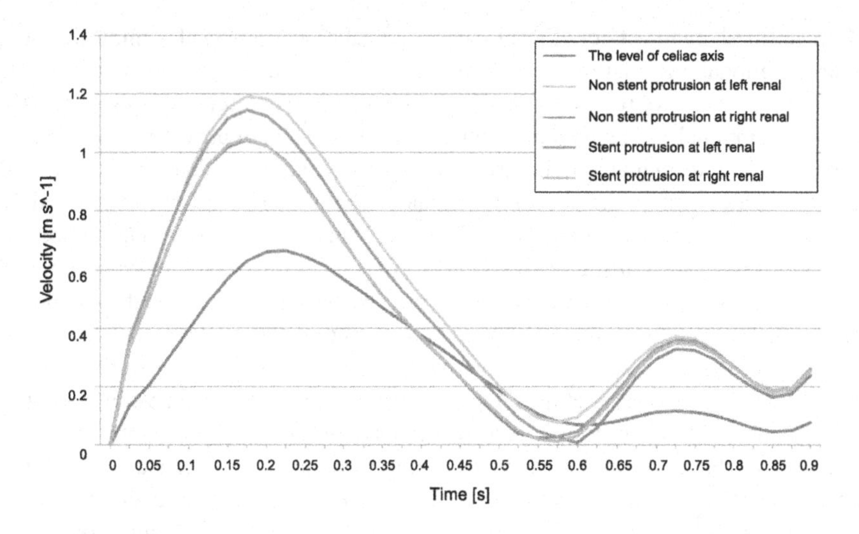

FIGURE 6.16
Flow velocity in a patient with simulation of fenestrated renal stents. Flow velocity is calculated after the placement of fenestrated stents at bilateral renal arteries in the realistic aorta models. Flow velocity is slightly decreased in the presence of stent protrusion (7 mm), although this does not result in the significant difference when compared to non-stent protrusion situation.

of fenestrated stent protrusion. With simulated fenestrated stents protruding into the abdominal aorta, flow recirculation patterns are observed in the proximal part of the renal arteries when compared to that seen in pre-operative graft implantation, although this does not result in any significant change of flow velocity (Fig. 6.16).[196]

6.6 Summary

Endovascular stent graft repair of AAAs has been regarded as an effective alternative to open surgery and this has been confirmed by many clinical trials. With the technical developments of stent graft devices, endovascular repair has been made available to more patients with aortic aneurysms, with satisfactory results having been achieved. Despite promising short- to mid-term results, long-term outcomes are still unknown, especially for suprarenal and fenestrated stent grafts. While clinical studies are required to conduct follow-up studies of patients from a long-term point of view with the inclusion of clinical assessment and imaging evaluation, the analysis of the haemodynamic effect of stent wires on renal blood flow is also important as it offers additional information when compared to traditional imaging evaluation.

In this chapter, we reviewed the device and technical evolution of stent grafts including suprarenal and fenestrated stent grafts with the aim of providing good understanding of the technical details associated with endovascular stent grafts. Medical imaging assessment of the successful implantation of stent grafts has been widely addressed in the literature and CT has been recognised as the preferred method for pre- and post-stent grafting evaluation. In order to improve our understanding of the long-term outcomes of endovascular repair, we have demonstrated the haemodynamic effect of stent wires on renal blood flow based on FSI analysis with the use of realistic aorta models. Although the analysis is limited to preliminary studies, it provides a basis for the testing of the effect of placing a suprarenal or fenestrated stent in front of/into the renal artery, and research findings provide insight into the treatment outcomes of these two modified stent grafting techniques. Further studies with the simulation of procedure-related complications such as endoleaks and measurements of the aneurysm sac pressure[197–201] are necessary to determine the long-term safety of suprarenal and fenestrated stent graft repair of AAAs.

6.7 Questions

1) List two recent technical developments proposed in endovascular repair mechanisms.
2) Explain the relationship between suprarenal and transrenal fixation of aortic stent grafts.
3) Explain the principles of fenestrated stent grafts.
4) Name three ways that can reveal the consequences of the suprarenal stent coverage of the renal arteries.
5) Name the equation used for computational modelling of blood flow.

7

Nasal Drug Delivery

CONTENTS

7.1 Review of Device 89
7.2 Computational Modelling 93
  7.2.1 Geometrical Meshing 93
  7.2.2 Physiological Boundary Conditions 95
  7.2.3 Simulating Flow in the Nasal Cavity 97
7.3 Assessment of Modelling and Optimisation 101
  7.3.1 Insertion Angle 101
  7.3.2 Full Spray Cone Angle 103
  7.3.3 Implications for Nasal Drug Delivery 107
7.4 Summary 107
7.5 Questions 110

7.1 Review of Device

There are many fluid flow applications in which two or more fluids exist with separate flow fields. All these types of flow are generically called multiphase, and they all can be simulated using the multiphase models in commonly available CFD tools using two approaches. Flow fields in some biomedical applications are complex and sometimes involve multi-phase flows. Multiphase flows are broadly classified into continuous–dispersed flow and continuous–continuous flow. Continuous–dispersed flow, as its name implies, is a model of a primary continuous fluid that has the secondary discontinuous phase dispersed within. A continuous–continuous flow model is one in which the inter-mixing fluids are continuous and with a free surface between the fluids. Typical CFD models for multiphase flows[202, 203] are (1) algebraic slip, (2) drift flux, (3) Eulerian–Lagrangian and (4) Eulerian–Eulerian. A judicious choice of multiphase models is the critical factor for an accurate CFD result.

Examining drug delivery performance within the anatomical environment is a good case of using multiphase flow simulation for analysis. The particle transport model is used to model flows with individual behaviour, and the Eulerian multiphase model is used for flows with overall information about a certain phase. It should be noted that the equations of multiphase flows are essentially those of single-phase flow applied to each fluid in the calculation, with two extensions. First, additional equations are included for the volume fraction of each fluid. Then, additional terms are included in the transport equations of each phase, to model the inter-phase transfers such as drag, heat transfer and mass transfer.

We present the CFD studies on nasal drug delivery that can be used to evaluate the performance of the spray devices. For this study, a computational model of the nasal cavity is reconstructed from CT scans, while the sprayed drug particles delivered from the nasal spray device can be simulated as a single release point or can include the spray device. After the reconstruction of the respiratory airways via medical imaging or optical scanning, computational techniques such as CFD can simulate critical information that are difficult to measure empirically, such as the air velocity profile,[204] particle trajectories and localised deposition sites.[205,206]

The model is prepared with a computational mesh and a commercial CFD software is used to run the numerical simulations. The numerical results provide details regarding airflow patterns, particle trajectories and the drug particle interactions with the inhaled air that leads to particle deposition and location. Such results offer a deeper insight into nasal drug delivery and are useful during product development cycles for the optimisation of the spray device.

While the CFD simulations provide critically important results, the accuracy and reliability of the results are highly dependent on setting good initial boundary conditions, among other prerequisites (e.g., numerical techniques, mesh quality, geometry reconstruction). The introduction of the sprayed particles into the nasal cavity airway requires the input of initial particle conditions. These may include the particle location, initial velocity and particle size. For this information, experimental measurements are needed to quantify the values. Some experimental work has been performed,[43,46,207] which provides some quantitative values for the initial particle conditions. In addition, experimental setups are also needed for the validation of the numerical setup. In these instances, the nasal cavity geometry and sprayed particles with or without the spray device need to be constructed physically and measurements can be made using PIV techniques, gamma scintigraphy imaging or scintillation counting.

The function of the nasal spray delivery device is to atomise the liquid formulation into a fine spray made up of small micron-sized particles. A number of deposition studies in the nasal cavity are dedicated to the inhalation of particles suspended in the airflow,[205,208–211] which is quite different to the inhalation of sprayed particles.[7,207] Figure 7.1 shows the different computational and experimental particle tracking techniques that have been used to examine drug inhalation characteristics.

In a step towards a more accurate representation of the nasal spray device, the inclusion of the nasal spray device as an object in the nasal cavity has been implemented and is demonstrated in this book. Figure 7.2 shows the effects of the inclusion

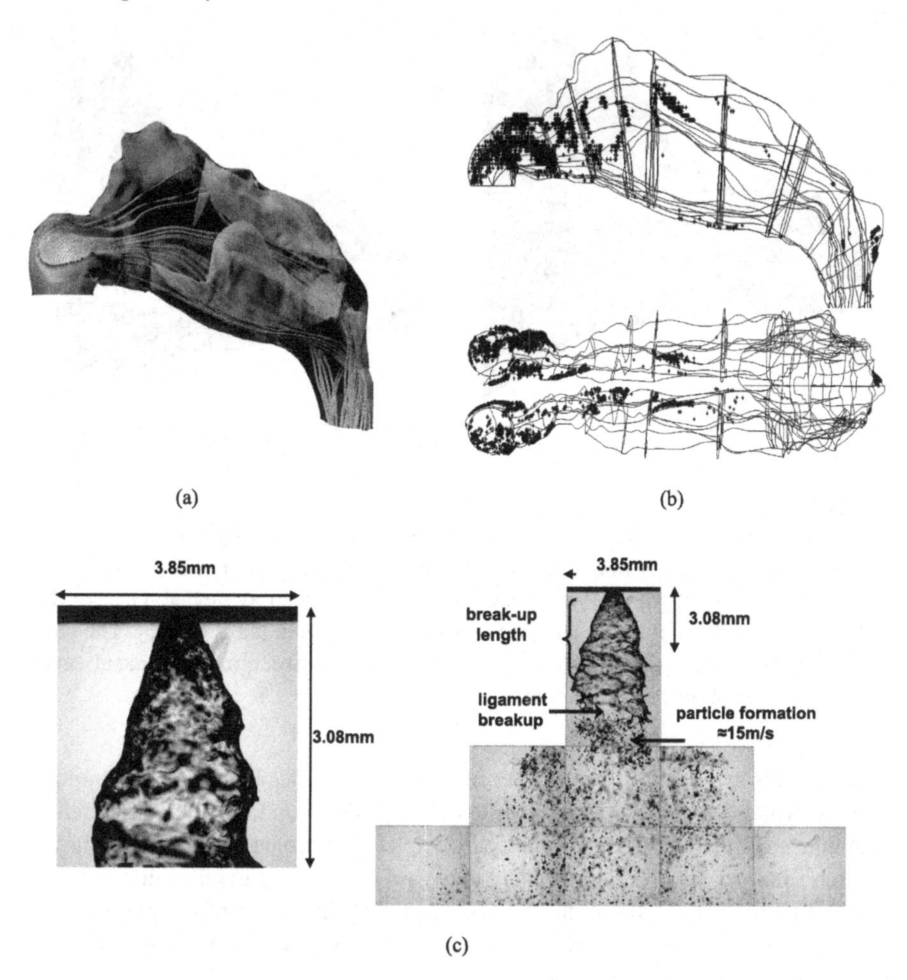

FIGURE 7.1
Nasal spray examination techniques. (a) Streamline tracing of inhalated suspended particles in air. (b) Particle tracking in the nasal cavity and externally to determine the origin of inhaled drug and its transport as delineated by the particle clouds. (c) Sprayed particle tracking showing particle release from a point source with the initial velocity and spray angle.

of the spray device on the airflow field, which is pertinent to the transport of the sprayed drug particles.

More recent advances in nanotechnology have seen the manufacture of engineered nanoparticles for many commercial and medical applications. Engineered

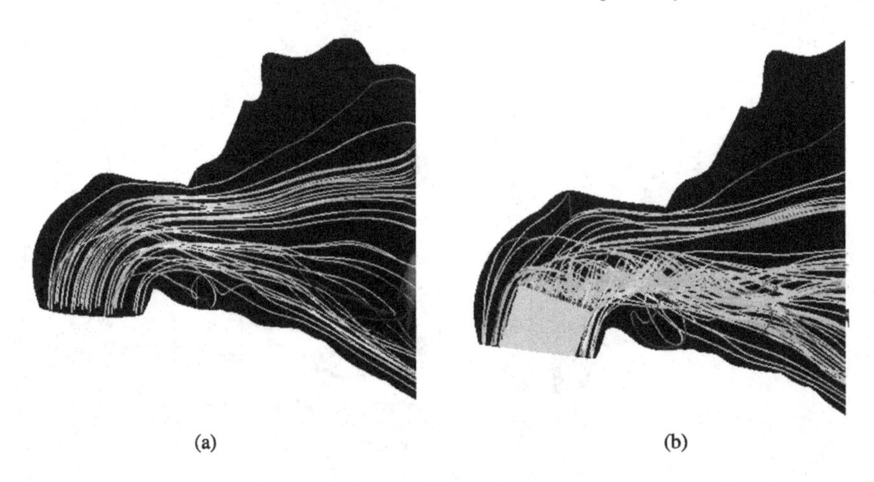

FIGURE 7.2
Simulated nasal spray performance. Path streamlines traced by massless particles and coloured by velocity magnitudes found in the anterior region of the nasal cavity, (a) without the nasal spray device and (b) with the nasal spray device. Greater disturbance is found when the nasal spray device is included in the geometry. The nasal valve region is displayed as it exhibits the smallest cross-section in the nasal cavity and hence, accelerates the streamlines.

nanoparticles can exhibit large surface area-to-size ratio leading to greater biologic activity, which can be desirable for drug delivery. This requires different modelling techniques compared with micron-sized particles, to account for Brownian motion of such small scales of particles being delivered through the nasal cavity.[205,212,213] The results in Fig. 7.3 shows the Brownian motion dispersion of the particles that diffuse through the nasal cavity, leading to a deposition pattern that is spread out in comparison to micron-sized particles.

The deposition pattern shown in Fig. 7.3 has interesting applications for drug delivery since traditional nasal sprays produce micron-sized droplets that are prone to inertial deposition. This deposition mechanism leads to high inertial impaction (up to 100% for the mean atomised particle droplet of 50 μm) in the anterior region of the nasal cavity.[7,207] However, for high drug efficacy, the delivery of the drug droplets needs to be deposited in the middle regions of the nasal cavity, where the highly vascularised walls exist. Smaller particles such as 1 μm were found to be less affected by inertial properties, which allowed it to bypass the anterior region of the nasal cavity. However, because of the particles' ability to follow the airflow streamlines more readily, the particles are less likely to deposit in any region of the nasal cavity at all and would bypass it completely, leading to the undesired effects of deep lung deposition. Therefore, the delivery of nanoparticles, especially 1-nm

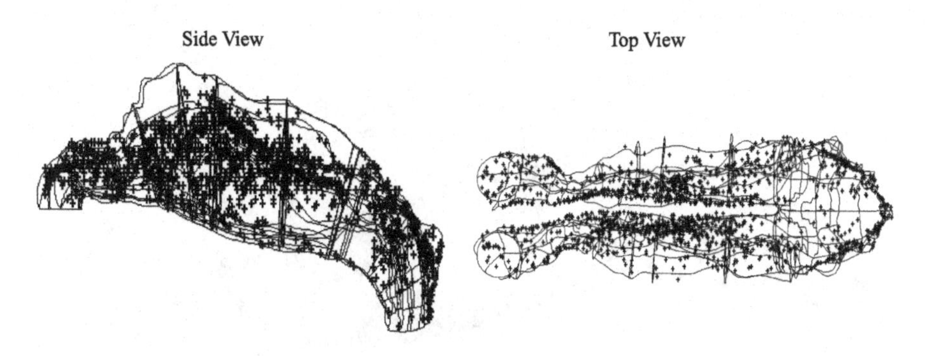

FIGURE 7.3
Deposition pattern of the drug by the nasal spray. Deposition patterns of 1-nm particles under a flow rate of 10 L/min in a human nasal cavity. The deposition efficiency is 81%, while the deposition pattern displays an even distribution throughout the geometry. This differs greatly from the deposition patterns for micron-sized particles which are usually concentrated to the anterior third of the cavity.

particles (81% deposition efficiency), can provide improved deposition in the middle regions whilst minimising deep lung deposition.

7.2 Computational Modelling

7.2.1 Geometrical Meshing

After extracting the nasal cavity airway by segmentation, the computational file should be saved as IGES, STL or STEP in order to be cross-compatible with a range of 3D modelling and meshing programs. Typically, the meshing procedure is begun by applying a simple unstructured tetrahedral mesh all over, which produces a single contiguous mesh. However, for easier postprocessing of local flow variables and particle deposition fractions, the computational model may be split into smaller sub regions during the CAD surface and volume generation stage, prior to meshing. While the process of sub-dividing the computational model into smaller regions can be performed within some CFD packages, it is not always an easy task and therefore, it is recommended to be performed in CAD packages that have NURBS functionality. Figure 7.4 shows a nasal cavity sub-divided into 10 sections.

Surfaces were then created and stitched up to create a computational mesh. An initial model with 82,000 cells was created and used to solve the airflow field at a flow rate of 10 L/min. The original model was refined by cell adaptation techniques

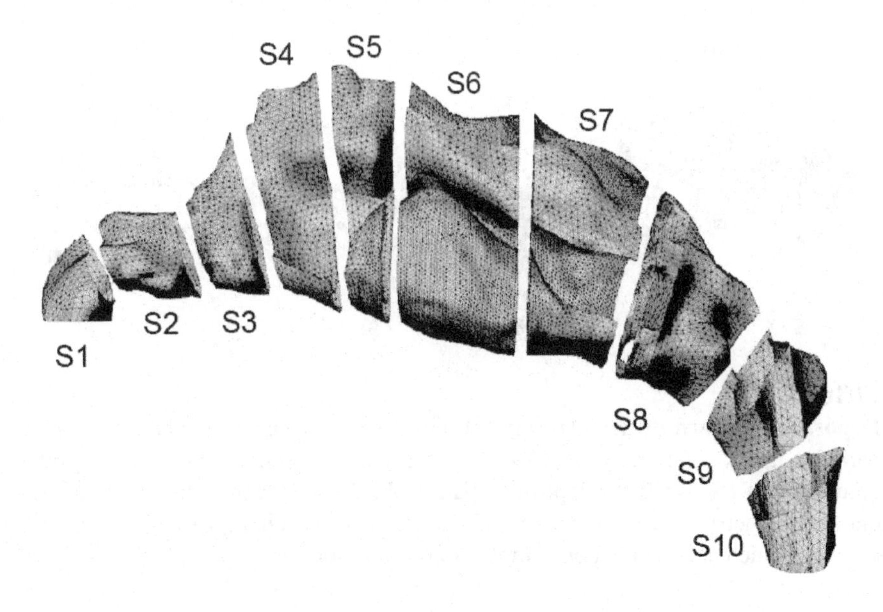

FIGURE 7.4
Nasal geometrical configuration. Nasal geometry subdivided into ten sections and labelled with S and a number representing the section. S1 to S3 represents the anterior, S4 to S7 the middle and S8 to S10 the posterior nasal cavity. An artificial extension is added after S10 to assist in the simulation.

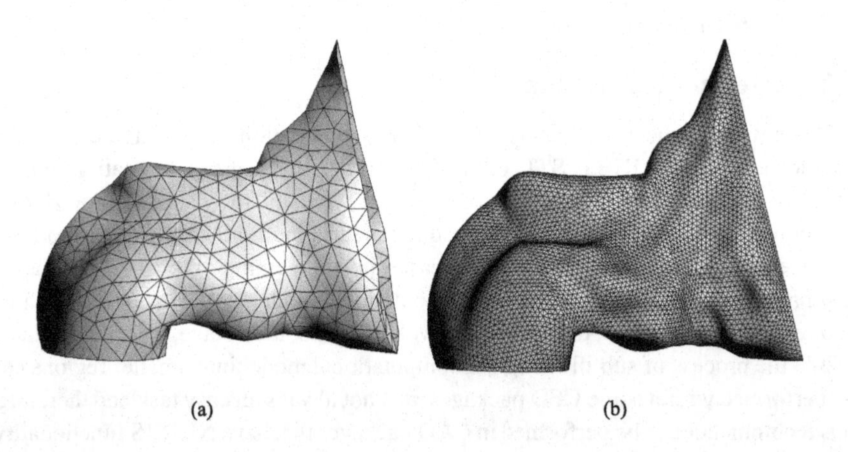

FIGURE 7.5
Nasal mesh configuration. Surface mesh of the first two sections S1 and S2 (anterior nose entrance region) from Fig. 4.19 of the nasal cavity. (a) An initial coarse mesh. (b) After mesh refinement, the mesh count is increased.

that included refining large volume cells, cells that displayed high velocity/pressure gradients and near-wall refinements. This process was repeated twice, with each repeat producing a model with a higher cell count than the previous model. A model containing 82,000 and 2.5 million cells are shown in Fig. 7.5.

Prism meshes are often used to resolve the thin boundary layers that are present in wall-bounded flows. The first mesh element, adjacent to the wall, is a very thin layer and subsequent mesh elements above the first mesh element become progressively thicker until the layers cover the distance of the boundary layer. As discussed earlier, the mesh should change slowly and smoothly away from the domain boundary (Fig. 7.6).

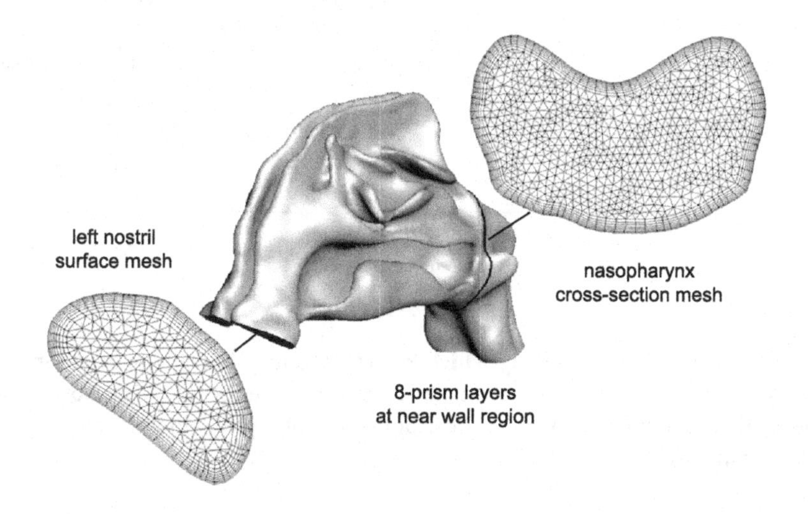

FIGURE 7.6
Near-wall mesh that has an eight-prism layer for a nasal cavity model. Application of a hybrid mesh that contains tetrahedral elements inside the domain and prism layers along the surface or domain boundaries.

7.2.2 Physiological Boundary Conditions

The reconstruction of anatomy and physiology of the nasal cavity by CAD has been discussed in Chapter 3. This is referred to again in Fig. 7.7. The nose can be divided axially into four regions — the vestibule, the nasal valve, the turbinate and the nasopharynx regions. The first three-quarters of the nasal cavity are divided into the left and right cavity separated by the nasal septum wall. Air enters each cavity through the oval-shaped external nostrils into the vestibule. The flow changes direction, 90° towards the horizontal, before entering the nasal valve region. The flow increases in

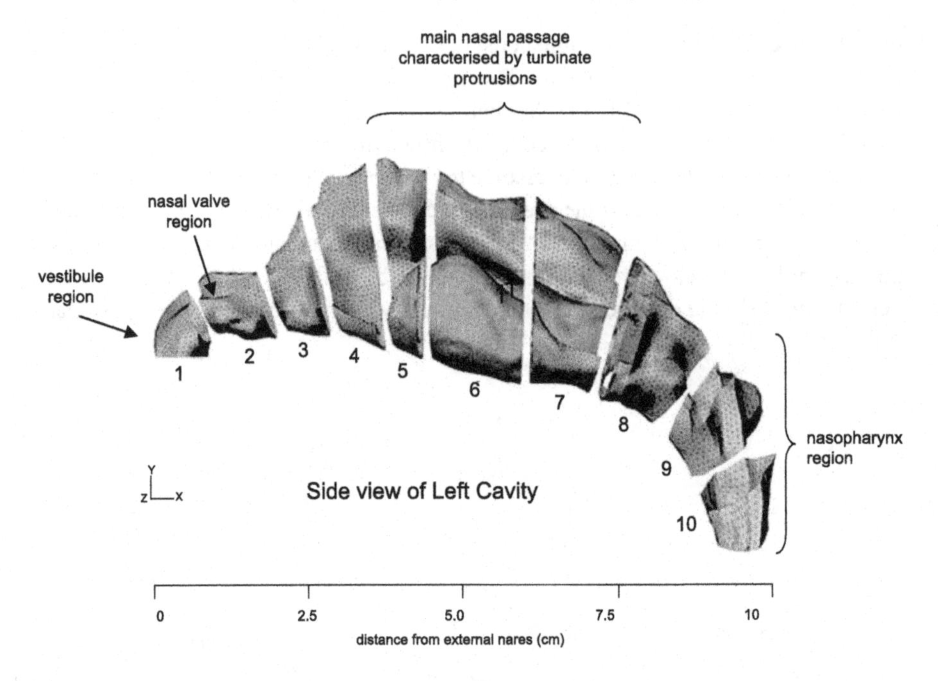

FIGURE 7.7
Side view of the left nasal cavity which is sectioned into 10 regions used for local deposition analysis. The $+x$ coordinate is from the anterior tip of the nostril inlet running posterior to the back of the nose at the nasopharynx, which is referred to as the axial direction.

this region where the cross-sectional area is smallest, causing an acceleration of the air. At the end of the nasal valve region, the cross-sectional area of the nasal cavity suddenly increases. This expansion is the beginning of the turbinate region where the profile is complicated and asymmetrical. Finally, at the nasopharyngeal region, the left and right cavities merge, causing the flow in this region to mix intensely.

The main factors that contribute to the airflow patterns are the nasal cavity geometry and the flow rate. For a realistic human nasal cavity, geometric variations between the left- and right-sided nasal cavities exist at any one time as a result of a combination of its natural geometry due to the nasal cycle and any other physiological reactions at the time. Inhalation is caused by the negative pressure induced by the diaphragm flattening out to increase the volume of the pleural cavity. Flow rates for adults can range between 5–12 L/min for light breathing and 12–40 L/min for non-normal conditions such as during exertion and physical exercise. Usually, breathing switches from pure nasal flow to oral–nasal flow at this higher range. Additionally, flow rates for extreme forced inhalation conditions have been found to reach 150 L/min.[214] Issues pertaining to the geometry and physiologic features also need to be kept in mind when setting up the numerical model and interpreting the results.

Laminar flows are normally characterised by a smooth motion where the fluid's viscosity dominates, allowing high molecular diffusion and dampening out any fluctuations in the flow. This leads to adjacent layers of fluid sliding past each other in an orderly fashion (like layers of lamina). However, for a nasal cavity which has a complex geometry that is highly convoluted, flow separation and recirculation can exist especially in the region from the nasal valve to the main nasal passage, where a rapid increase in the cross-sectional area is observed,[215] enhancing flow instabilities. It must be mentioned that although flow separation and recirculation are typical characteristics of turbulent flow, the presence of these characteristics does not necessarily assume that the flow is turbulent, since its existence can be found in laminar flows with geometries that exhibit separation (e.g., flow over a backward-facing step or a cylinder).

Boundary conditions for the computational surfaces need to be defined. While the surface walls are easily understood in terms of its definition as a rigid wall boundary, the nostril inlets and the nasopharynx outlet provide the user with more options for definition. In this case study, a uniform flow perpendicular to the nostril inlet was specified. This assumption was based on the data of Keyhani *et al.*,[216] who demonstrated that a velocity profile at the nostrils for a given flow rate did not show significant differences on the downstream flow field when compared with experimental data. In addition, the flow rates of left and right nostrils are assumed to be the same. This does not simulate real breathing perfectly since the flow is triggered from the lungs, drawing the air from the nostrils, which are affected by geometrical differences, leading to varied flow rates between the cavities. The breathing cycle is caused by the pressure difference induced by the diaphragm flattening/contracting to increase/decrease the volume of the pleural cavity. Therefore, it is natural to assign the nostril inlets and the nasopharynx outlet as pressure conditions.

7.2.3 Simulating Flow in the Nasal Cavity

Putting all these steps together, we see that the CFD solution process culminates in the processing and analysis of the numerical data that is produced. In its raw form, this data is simply a collection of numbers stored at coordinates based on the mesh design. The conversion of raw data into meaningful results is called the post-processing stage. CFD has the capability to produce colourful graphics and precise detailed data. In this section, we present some essential postprocessing techniques frequently encountered in the presentation of CFD data.

Streamlines are used for examining the nature of a flow either in two or three dimensions (e.g., Fig. 7.8). By definition, streamlines are parallel to the mean velocity

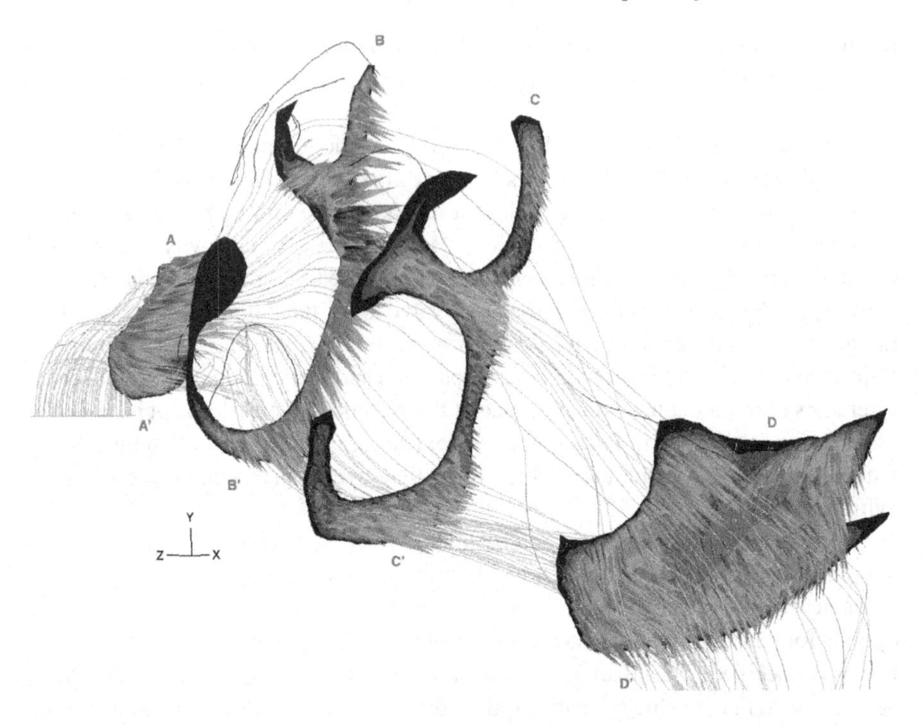

FIGURE 7.8
Streamline traces of airflow within a nasal cavity model. Airflow streamlines passing through velocity contours taken at different coronal sections in the right nasal cavity. Velocity vectors on each coronal section shaped by cones are also presented.

vector, where they trace the flow pattern using massless particles. They can generally be obtained by integrating the spatial three velocity components expressed in a 3D Cartesian frame: $dx/dt = u$, $dy/dt = v$ and $dz/dt = w$. Streamlines are also called streamtraces, streaklines or path lines.

The effects of swirl can be investigated by setting the swirl fraction. For this demonstration, a flow rate of 10 L/min laminar flow is used. The swirl fraction sets the fraction of the velocity magnitude to go into the swirling component of the flow, thus a higher fraction will produce a greater tangential velocity component. This increases the time taken to travel a given axial distance as the particle is swirling more and its residence time becomes longer. Additionally, the induced drag from the cross-flow of air helps to reduce the initial high momentum of the particle and the chance of particles travelling through the frontal regions of the nasal cavity increases. Figure 7.9 shows the high deposition of particles occurring in the frontal regions (Sections 1–3).

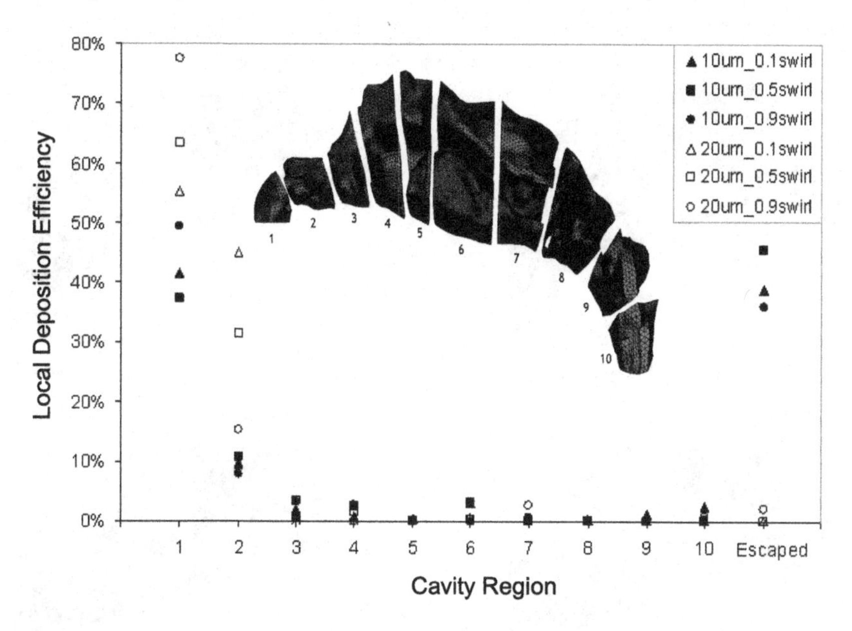

FIGURE 7.9
Particle deposition within different sections of the nasal cavity. Sections 1–3 are the frontal region, Sections 4–7 are the middle region and Sections 8–10 are the posterior region.

The amount of swirl has a different effect for the two different particle sizes. For 10-μm particles with $\lambda = 0.9$, deposition in the frontal zones increases, while for $\lambda = 0.5$, a higher percentage of particles escape. Conversely for 20-μm particles, an increase in the swirl fraction decreases the deposition in the frontal zones. Some deposition in the middle zones occurs and 2.2% escape when $\lambda = 0.9$. The reason for these local deposition patterns can be better understood by observing the particle trajectories (Fig. 7.10). It is known that larger particles that possess high inertia (i.e., Stokes number characteristics) need to be aligned with the flow streamlines to avoid impaction. This implies that the particles need to be projected at a clear unobstructed path in the nasal airway rather than be projected at walls.

When $\lambda \to 0$, the velocity magnitude is entirely composed of axial velocity, which projects the particles vertically. High inertia particles are directed at the roof of the nasal vestibule and do not have enough time to slow down and adapt to the gas phase streamlines. However, smaller inertia particles can adapt to the flow streamlines more readily. One idea to overcome this is to insert the nasal spray at an angle (insertion angle) that would provide such alignments with the flow streamlines.[7] However, this technique is dependent on the user upon the application of the

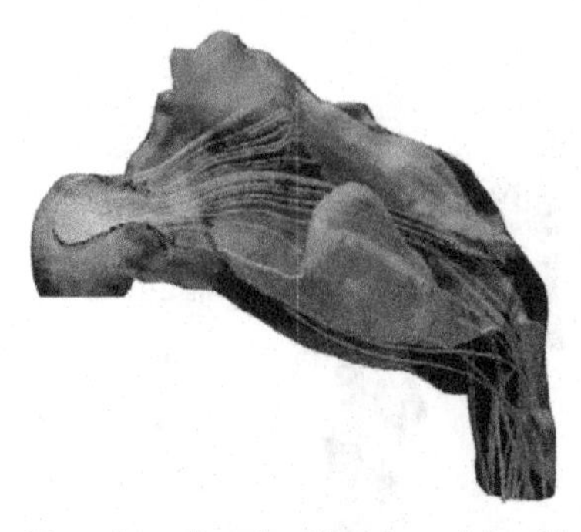

10μm, 4mm diameter at break-up, 0.1 swirl

20μm, 4mm diameter at break-up, 0.1 swirl

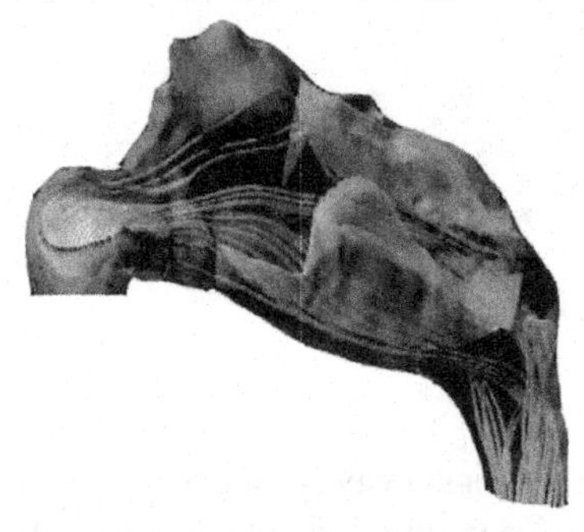

10μm, 4mm diameter at break-up, 0.5 swirl

20μm, 4mm diameter at break-up, 0.5 swirl

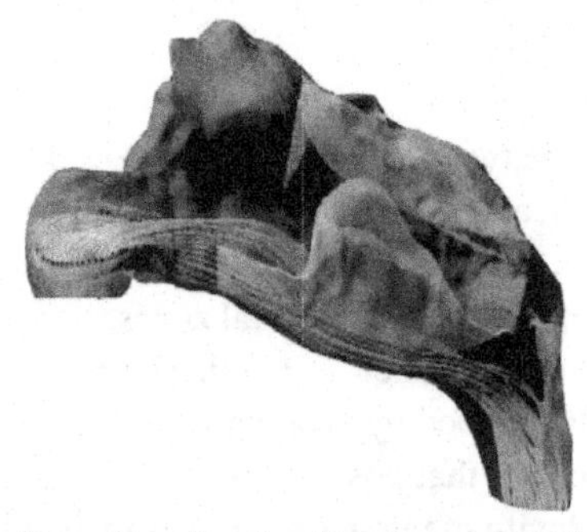

10μm, 4mm diameter at break-up, 0.9 swirl

20μm, 4mm diameter at break-up, 0.9 swirl

0.00e+00 1.09e+00 2.18e+00

FIGURE 7.10
Particle trajectories within the nasal cavity for different swirls. Particle trajectories for 10- and 20-μm particles at different swirl fractions are presented for 0.1 to 0.9 swirls.

nasal spray device as well as the precise location of the nozzle. When $\lambda \to 1$, the velocity magnitude is entirely composed of a tangential velocity and the particles are projected at a tangential direction to the nasal spray insertion angle. For 10-μm particles, deposition increases in the frontal zones as more particles are pushed towards the wall. For 20-μm particles, this is also the case; however, instead of impaction into the upper roof when $\lambda = 0.1$, impaction is on the side walls. The horizontal projection of a small proportion of particles is then enhanced by the gas phase velocity, which carries the particles along the floor of the nasal cavity.

Without any obstructions to the flow, the particles are transported through to the nasopharynx area. A small increase in deposition within the middle and anterior regions are observed. At the nasopharynx, the flow changes direction at 90° which acts as another filter to capture high inertia particles that impact at the back. Although a swirl fraction of 1 is extreme and indeed unrealistic to achieve practically within the atomiser design, its use in the present investigation provides a better understanding of the effects of the swirl fraction. The particle trajectories demonstrate that the swirling motion decreases the particles' initially high inertia due to the difference in the air and particle velocities.

7.3 Assessment of Modelling and Optimisation

7.3.1 Insertion Angle

The insertion angle, α, is the angle between the nasal spray device from the horizontal position, (0° in the x–y plane), when looking into the side of a person's face (Fig. 7.11). Particle sizes of 10 μm to 20 μm were used at an initial velocity of 10 m/s. A uniform surface injection released from the inlet was used to eliminate variables such as location of the nozzle tip, nozzle diameter and spray cone angle.

Higher deposition occurred at 100° for most particles (Fig. 7.12). This concentration of particles can be seen in Fig. 7.13 for 15-μm particles. Minimum deposition for smaller-sized particles, 10 μm and 15 μm, was found when $\alpha = 70°$. The 70° direction of particles enhances the ability of turning, as more particles assume the streamlines on the inside of the curvature. Further decreases in α, where the direction of discharge approaches the horizontal, increases the deposition of particles as a higher proportion of particles is now directed at the adjoining wall of the nostril, albeit at a small margin.

The 20-μm particles behave similarly, but at lower α. Minimum deposition in the two frontal zones was found at $\alpha = 45°$. Although more particles are directed into the adjoining wall, this is offset by more particles penetrating into the curvature instead of impacting straight into the roof of the vestibule as is the case when $\alpha = 90°$. These

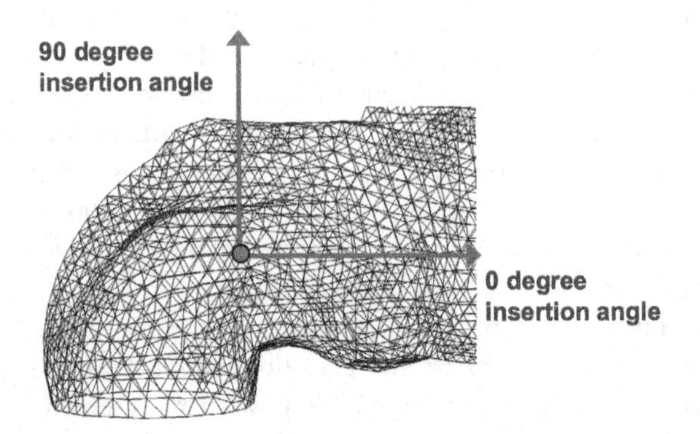

FIGURE 7.11
Side view of the nasal vestibule showing the insertion angle. This diagram illustrates the insertion angle α, of the nasal spray device from the side view of a face of a human subject.

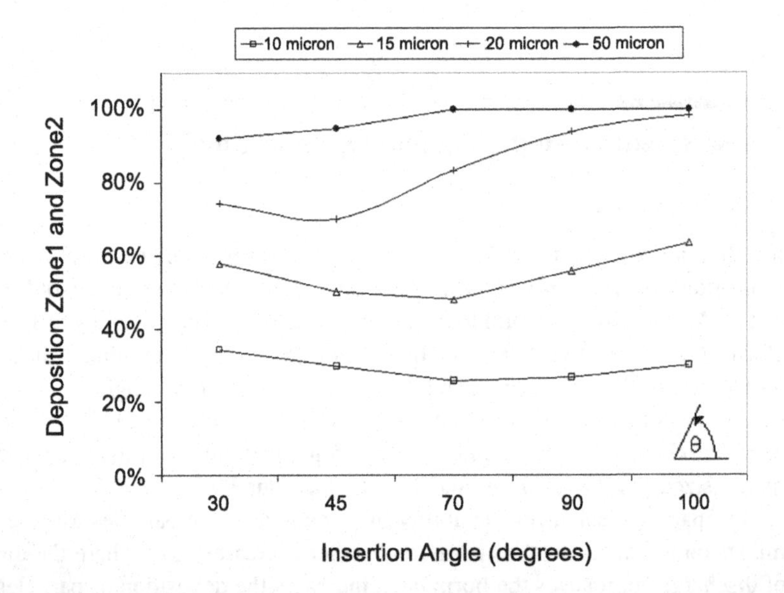

FIGURE 7.12
Deposition in Zone1 and Zone2 for monodispersed particles released uniformly from the inlet surface at different insertion angles. As the gas phase velocity cannot overcome the initial injected velocity of 10 m/s, the particles discharged directly into the anterior-most walls of the nasal cavity.

larger particles require a sharper angle of insertion to avoid impacting with the roof of the vestibule, thus aiding the alignment of the particles with the 90° bend and reducing the amount of deviation required in turning. The larger the particle (20 μm and 50 μm), the more effective the decrease of α becomes. Another insertion angle that can be considered is the orientation when looking into a person's face, front-on, in the y–z plane. This was not investigated as the same ideas regarding the particle size with its dependency on initial flow conditions exist.

7.3.2 Full Spray Cone Angle

The full spray cone angle β, is the dispersion of particles exiting from the nozzle tip (Fig. 7.14). Particles were released from a small diameter compared with previous particle release methods, which used a uniform surface inlet at the nostril openings. This allows observation of the physical differences when changing β. Particles were released at 10 m/s from an internally fixed location with a diameter of 0.8 mm and a range of β between 20° to 80°.

The deposition of 10-μm particles in the two frontal zones are unaffected by the change in β (Fig. 7.15). Again, the significance of β is realised as particle size increases. The smaller ranged particles, that follow the gas phase velocity more readily, are optimised when released with a narrow β. A wider β gives rise to a larger range of dispersion due to the nature of a 360° spray cone.

There is a low ratio of favourably dispersed particles (those pointing with the flow) to those being dispersed away from the curvature of the gas flow. Figure 7.16(a) shows the flow for 15 μm being centralised when $\beta = 20°$ and the increase in deviation from the centre when $\beta = 80°$. Deposition for $\beta = 20°$ remains along the roof of the nasal cavity and near the septum walls with 33% depositing in the first two zones. At $\beta = 80°$, a higher proportion of deposition is observed in the frontal zones.

The internal location of injection is closer to the roof of the vestibule than from the nostril inlet. This reduces the allowable distance for particles to relax their initial injected conditions to the gas phase conditions, thus enhancing impaction on the roof of the vestibule. This effect is more pronounced for larger particles, i.e., 20–50 μm. As seen earlier, these particles have near 100% deposition in the front two zones. Consequently, any deviation that is favourable will project the particles into the already curved streamlines, allowing particles to travel further, albeit a small proportion. Figure 7.16(b) compares the two deposition patterns for 20-μm particles at $\beta = 20°$ and $\beta = 80°$. At $\beta = 20°$, impaction occurs directly above the injection release point in a concentrated area. When $\beta = 80°$, a wider area of deposition is observed in the frontal zones, whilst those particles projected favourably towards the nasal valve are able to travel beyond the 90° bend. However, their deposition is imminent and this occurs within the middle sections of the nasal cavities. Particles in the 15–20 μm range will exhibit an optimum β, which is based on the ratio of the change

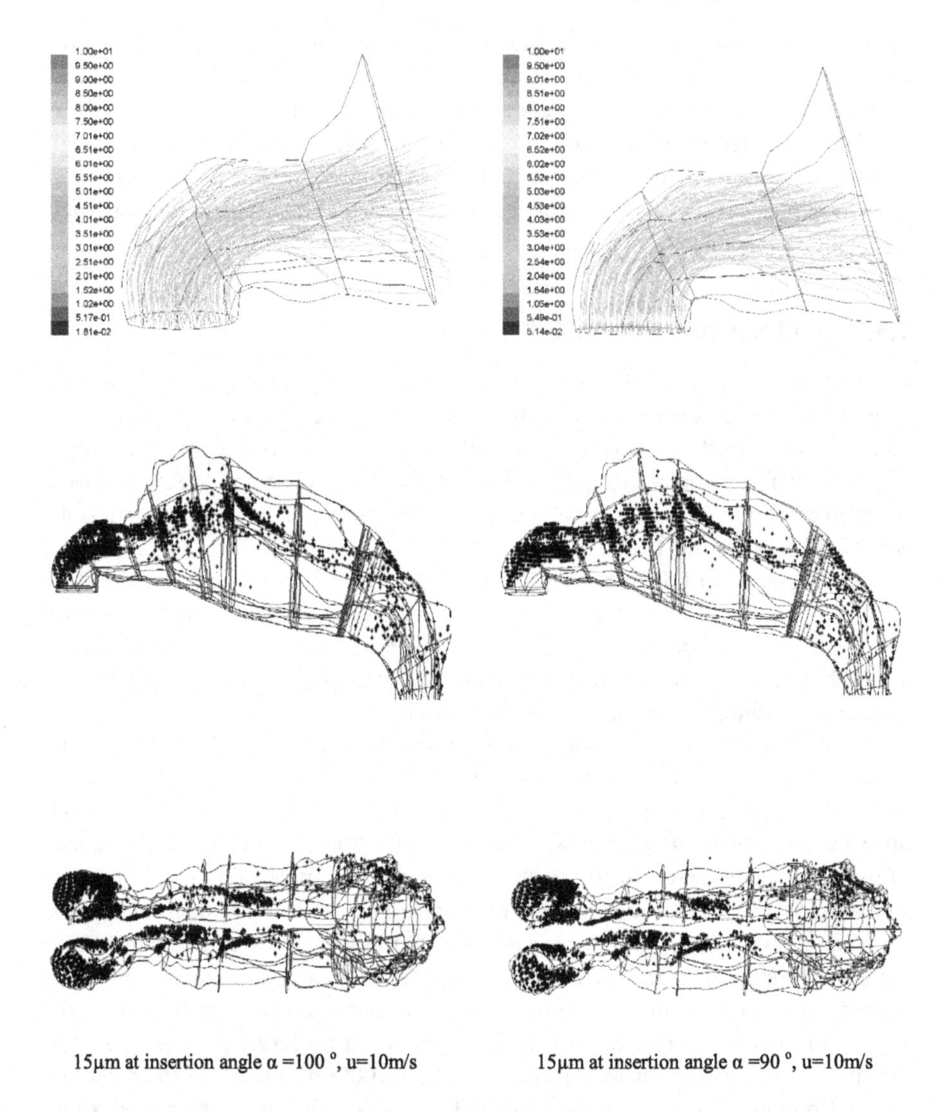

(a)

FIGURE 7.13

Deposition based on different particle sizes and insertion angles. (a) Deposition patterns for 15-µm particles released uniformly from the inlet surface uniformly from the inlet surface at insertion angles 100° and 90°. (b) Deposition patterns for 15-µm particles released uniformly from the inlet surface uniformly from the inlet surface at insertion angles 70° and 30°.

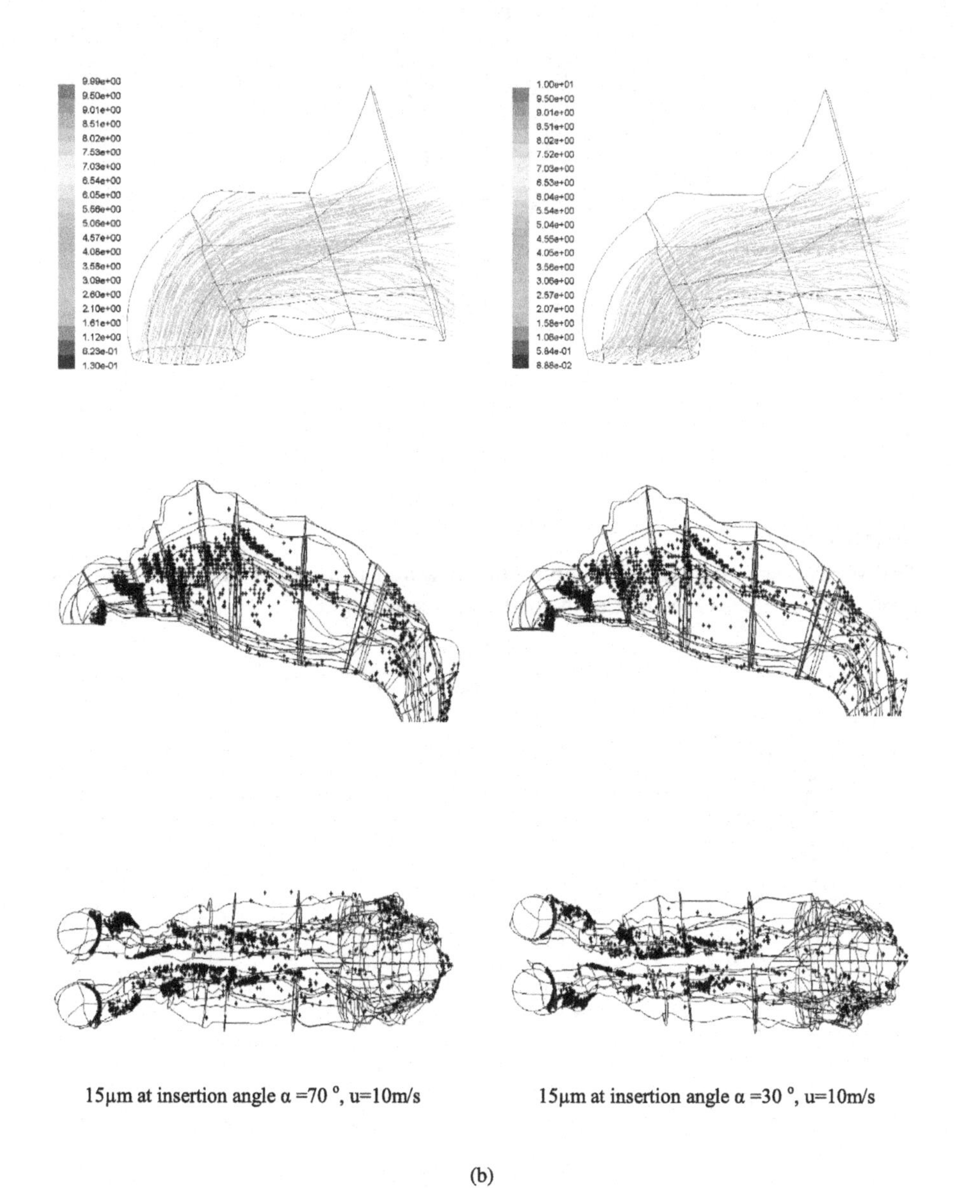

15μm at insertion angle α =70 °, u=10m/s

15μm at insertion angle α =30 °, u=10m/s

(b)

FIGURE 7.13
(Continued).

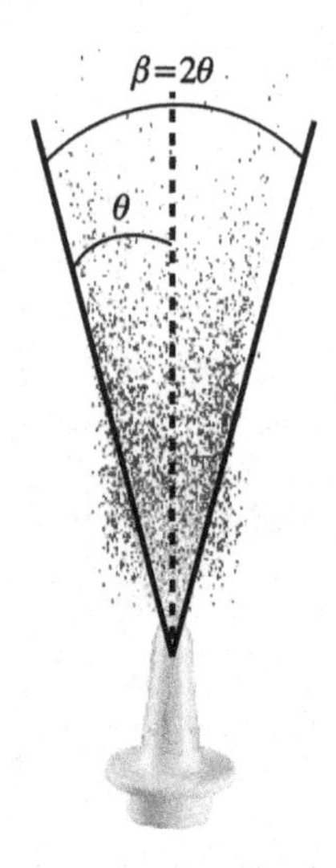

FIGURE 7.14
Full spray cone angle (β) in comparison with half cone spray angle (θ). The full spray cone angle is twice that of the half spray cone angle.

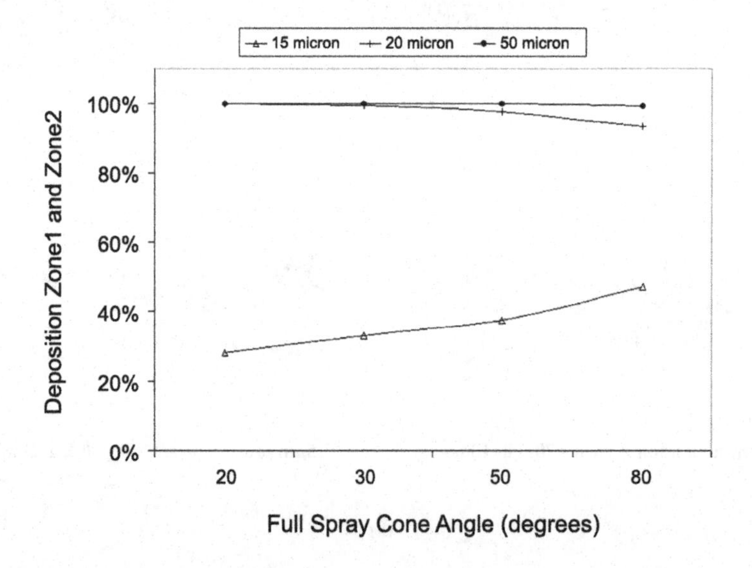

FIGURE 7.15
Variation of particle deposition patterns based on spray cone angles. Deposition in Zone1 and Zone2 for monodispersed particles released at 10 m/s from a small internal diameter at different spray cone angles.

in favourably dispersed particles to the number of particles that are predestined to impact on the roof of the vestibule because of their particle size.

7.3.3 Implications for Nasal Drug Delivery

The results from the case study were aimed at gaining a greater insight into which parameters are important in the design of nasal drug delivery devices. The results revealed high deposition occurring in the frontal regions of the nasal cavity. It is well recognised that one of the functions of the nose is to filter out foreign particles during inhalation, which was mainly thought to be attributed to cilia (nasal hairs) movement within the nose. However, the filtering function also extends to the airway geometry at multiple locations in the nasal cavity. In the frontal sections (Regions 1 and 2), the flow experiences a curvature in the streamlines which acts similarly to inertial impactors that filter out high inertial, large particles. Another filter is the contraction of the nasal airway into the smallest cross-sectional area, the nasal valve region where particles are accelerated increasing their inertial property. Finally, another right angle bend at the nasopharynx serves as a final filter prior to the particles entering the lower airways. This may be one explanation for the undesirable "taste" patients experience after taking drugs via the nasal cavity.

The filtering curvature in the frontal sections along with the constricting nasal valve region is most significant for therapeutic drug delivery as it prohibits larger particles from penetrating into the middle cavity region for deposition onto the highly vascularised mucosal walls. One possible solution is to instruct users to insert the spray deeper or to develop spray devices that will naturally be inserted deeper with particles being released beyond the frontal curvature.

7.4 Summary

The case study presented here investigated the application of nasal drug delivery. While the nasal cavity's natural function is to filter out toxic particles, there is an opportunity to exploit the highly vascularised walls for drug delivery. Nasal sprays deliver atomised particles into the nasal cavity at high velocities. The structure of the atomised spray from a nasal spray was presented in order to identify the parameters that would affect the setting up of the initial particle boundary conditions for the CFD simulations. Parameters that were important, among others, included swirl effects, the spray cone angle, initial particle velocity and the insertion angle, which were evaluated for deposition efficiency under high and low inertial particles.

For a flow rate of 20 L/min, the 10–20-μm particles were sensitive to initial injection velocity, insertion angle and spray cone angle as their size increased. Larger

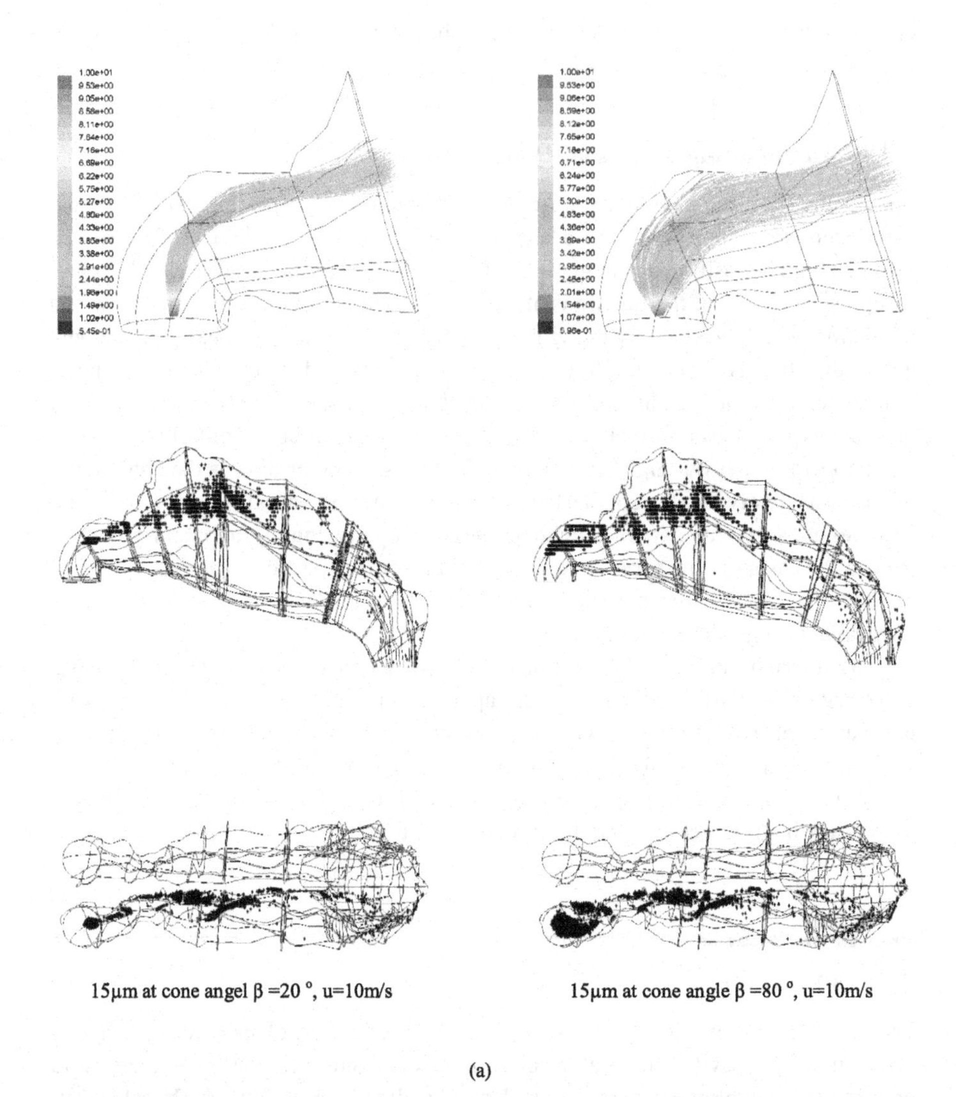

(a)

FIGURE 7.16

Particle deposition patterns based on particle sizes and spray cone angles. (a) Deposition patterns for 15-μm particles released at 10 m/s from a small internal diameter at the centre of the nostril inlet surface. The spray cone angles ranged from 20° and 80°. (b) Deposition patterns for 20-μm particles released at 10 m/s from a small internal diameter at the centre of the nostril inlet surface. The spray cone angles ranged from 20° and 80°.

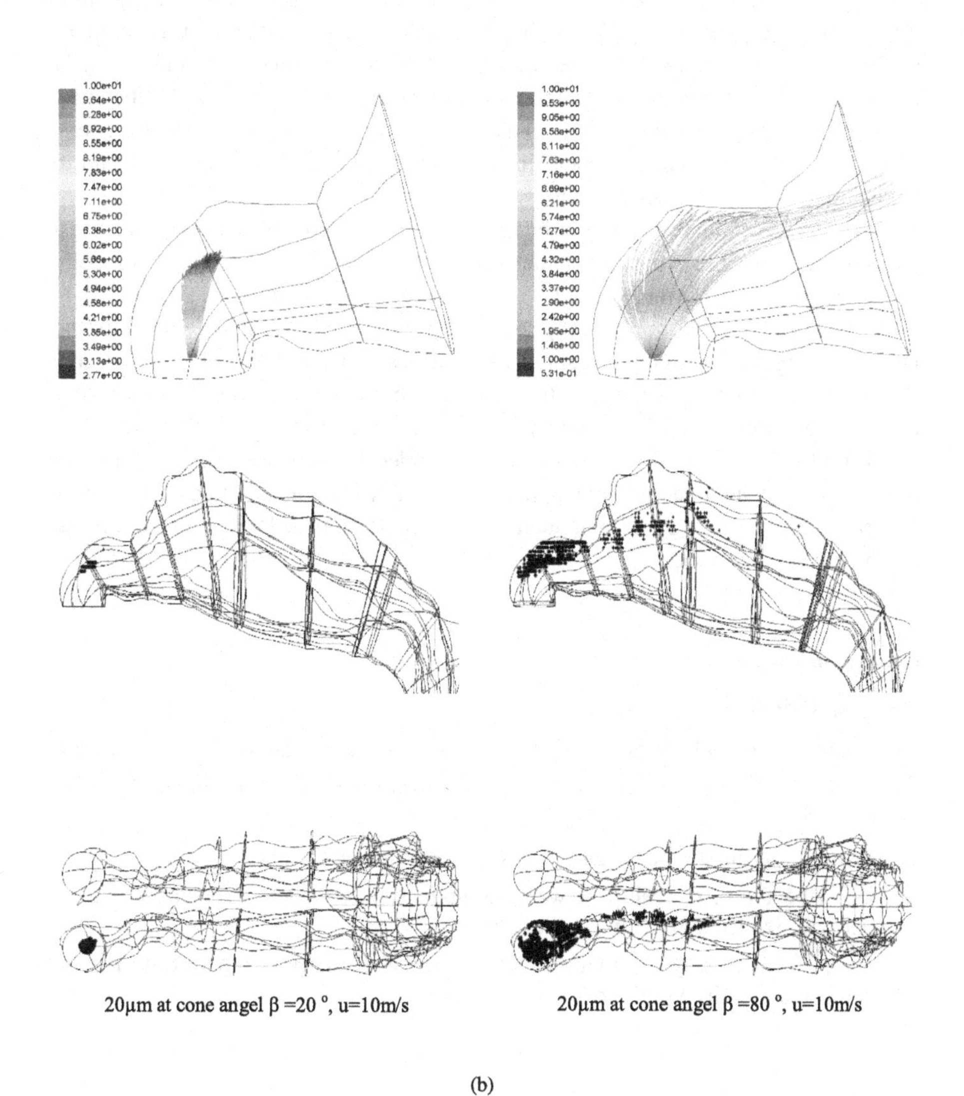

FIGURE 7.16
(Continued).

particles exhibited very high Stokes numbers (inertial parameter) causing it to be insensitive to these parameters. Current commercially available nasal sprays produce mean size particles of 45–60 μm. This presents a problem as larger particles (≥ 20 μm) are relatively insensitive towards initial injection conditions and are likely to deposit in the anterior portion of the nose, decreasing the drug delivery's efficiency. Producing smaller particles (≥ 20 μm) during atomisation is an option for designers; however, smaller particles are more inclined to follow the airflow and can lead to deposition beyond the nasal pharynx.

The ideas formulated can be used as a basis for improving the design of nasal spray devices to achieve better drug delivery such as (1) redirecting the release point of the spray (i.e., the insertion angle) to align with flow streamlines, (2) controlling the particle size distribution and (3) controlling the particle's initial velocity. In the attempt to replicate actual nasal spray applications while isolating the investigated parameters, idealised injected conditions for the particles were used. Further studies have been performed,[40,217] which also tested some of the spray characteristics. Interestingly, this field of research can be extended to compare results with other nasal cavities to include the permeability effects of nasal hairs and to establish more accurate initial particle conditions such as the instantaneous velocity at injection that can include swirl effects.

7.5 Questions

1) Name four modelling factors that define accurate and reliable CFD simulations.
2) What are the main factors that contribute towards airflow patterns in the nasal cavity?
3) What is the function of a nasal spray delivery device?
4) Describe the relationship between monodispersed particle deposition and the particle size for a given insertion angle.
5) Describe three ways that can used to improve nasal spray devices in order to achieve better drug delivery.

8

Biomedical MEMS Micropump

CONTENTS

8.1 Review of Device 111
8.2 Biomedical Applications of MEMS Micropumps 112
8.2.1 Potential Drug Delivery Applications 112
8.2.2 Other Potential Applications 113
8.3 Numerical Modelling of Micropumps 114
8.4 Operating Principle of an Example Micropump 115
8.5 Theoretical Analysis 115
8.5.1 Actuation Force 117
8.5.2 Microfluidic Pressure Variation 117
8.6 Model Development of an Example Micropump 118
8.7 Modelling and Simulation 118
8.8 FEA-based CFD Simulations 121
8.8.1 Inlet and Outlet Flow Characteristics 123
8.8.2 Flow Rate Estimation 124
8.9 Summary 125
8.10 Questions 126

8.1 Review of Device

Due to the rapid growth in MEMS technology, the design, development and realisation of miniature devices for biomedical implants can form a unique development philosophy. In particular, Bio-MEMS–based implantable micro-drug delivery systems have shown the potential to offer new paradigms in biomedicine and biology. These systems consist of various types of MEMS devices such as micropumps, microsensors, microvalves, microneedles, microfluidic channels and drug reservoirs.[38,218,219]

A typical micropump is a fundamental and a critical part of a drug delivery system, providing an actuation source to effectively transfer an accurate amount of fluid or drug to the targeted location. Among the large number of microfluidic components realised up to now, micropumps have clearly represented a critical role in science.[38,220,221] Such a system is mainly aimed at abrupt life threats such as heart attack, stroke, septicaemia or serious chronic diseases such as diabetes, melancholia and malignant lymphoma. With an automatic dosing system being active, sudden death could be prevented, in addition to reducing the risk of irregular or incorrect intake of medicine and drugs. In recent decades, the advancement of medical imaging, computer technology and manufacturing processes have added value to the conceptualisation and development of new biomedical applications, and significantly improved health and living standards, as well as contributing to medical science and technology.

Furthermore, in the design and development of such microfluidic devices, there has been a rapid introduction and utilisation of computational modelling tools for design level parameter optimisation and optimal performance analysis. Computational tools such as CAD software and numerical solvers for FEA and CFD packages are mainly used in modelling, simulation, analysis and verification stages of the development.

FEA is classified under computational simulation techniques that is able to evaluate structural performance of the device. Considering the ability to model more advanced geometries, FEA has become a popular computational numerical method, as opposed to theoretical or analytical modelling. Various researchers have utilised advanced FEA techniques for the modelling and analysis of implantable devices for applications such as drug delivery systems.[222]

CFD is a numerical method that can be successfully utilised to perform detailed flow analysis in microfluidic devices.[21,22] Furthermore, CFD and FEA techniques together facilitate the validation of results against experiments and are highly regarded in modern device design. Once validated, the model can be used to investigate the effects of changing parameters or geometry with greater certainty, and at substantially less cost than building a new experimental prototype. In addition, the simulation of design performance can be achieved to evaluate the safety of the device implant or the effect of introducing drugs into a system without endangering human lives. It may serve to provide expert opinion to surgeons in the event of strategising the device implant through better understanding of its operating mechanism.

More importantly, the targeted applications of the device also need to be considered during the design stage, in order to improve the suitability of such devices for desired applications. The next section in this chapter briefly illustrates some of the potential biomedical applications of MEMS micropumps.

8.2 Biomedical Applications of MEMS Micropumps

8.2.1 Potential Drug Delivery Applications

Biomedical micropumps can be utilised for the site-specific drug dosing for various applications such as hyptertension control, cancer treatments and diabetes control.

TABLE 8.1
Commonly used anti-hypertensive drugs. These medications mainly block specific type of hormone receptors to control hypertension, hence a range of other disorders.[37]

Anti-hypertensive medication	Effect
Beta-blockers	Block the effects of adrenaline and ease the heart's pumping action.
Angiotensin-converting enzyme (ACE)	Reduce the production of angiotensin, which is a inhibitors chemical that causes arteries to narrow.
Angiotensin receptor blockers (ARBs)	Block angiotensin, which is another chemical that causes arteries to narrow.
Calcium channel blockers (CCBs)	Decrease the contractions of the heart and widen blood vessels.

Here, a brief analysis is presented, highlighting the potential use of micropumps in controlling hypertension.

Drug Delivery for Hypertension: High blood pressure, also called hypertension, is elevated blood pressure in the arteries. Hypertension results from two major factors, which can be presented independently or combined: (1) the heart pumps blood with excessive force and (2) the body's smaller blood vessels (arterioles) become narrow, so that blood flow exerts more pressure against the vessel walls.

It has been reported that hypertension places stress on several organs (target organs), including the kidneys, eyes and heart, by causing them to deteriorate in functionality over time. Furthermore, it has been demonstrated that high blood pressure contributes to 75% of all strokes and heart attacks.[37]

In addition to lifestyle and dietary changes, various types of drugs are available as a major treatment for hypertension. Some of the commonly used anti-hypertensive medications are listed in Table 8.1. In some situations, the targeted quick delivery of the correct amount of drug would make a difference between life and death for a patient.

Therefore, based on the various novel technologies and features associated with modern mechanical micropumps, such devices could be implemented to facilitate the targeted and effective delivery of commonly used anti-hypertensive drugs, especially those that block specific types of hormone receptors.

8.2.2 Other Potential Applications

In addition to drug delivery, MEMS micropumps complemented with wireless and secure communication technology could be implemented in a range of other applications including contraception for men, ultrafast DNA sequencing, flow cytometry, LoC applications, micromixing, as well as non-biomedical applications such as fuel cell design, structural health monitoring and ink jet printer head modelling. Here, a brief illustration is presented, highlighting the implementation of a wirelessly controlled SAW-based secure microactuation principle in relation to contraception for men.

A wirelessly controlled and securely operated biomedical micropump can be extended to introduce a novel and innovative approach to prevent unintended pregnancy, by implementing a non-hormonal contraceptive method for men, by means of implanting a SAW device-based, secure microvalve through the vas deferens, which interferes and prevents sperm delivery. Such a microvalve would incorporate highly relevant features such as wireless control, battery-less operation, small size and mass, biocompatibility with no active electronics, low cost fabrication and increased reliability.

One of the key novelties of such a contraceptive device is that coded RF energy is transduced directly to perform mechanical actuation, only when the implanted code matches the code in the interrogation signal. Therefore, this innovation can successfully eliminate the requirement of a battery and active electronics to go with the micropump. Hence, the micropump can be substantially miniaturised to implant through the vas deferens, without any requirement of subsequent surgery to replace power sources.

Moreover, because of the ability to design such micropumps in an easily and safely reversible manner, the methodology is immune to common side effects such as the generation of enhanced antibodies that destroy sperms and the reduction of fertility due to back pressure caused by permanent blocking that exist in current methods.[223]

Once an exposure to a potential application domain is achieved, the numerical modelling of such micropumps can be performed with the utilisation of various tools and techniques as discussed.

8.3 Numerical Modelling of Micropumps

Advanced numerical modelling and FSI analysis between multiple physical fields of micropumps is of great advantage especially in optimising critical device parameters prior to fabrication. Many researchers have investigated the numerical modelling of electromechanical fluid coupling in micropumps.[224,225] However, most of these modelling has been on piezoelectric micropumps and the analyses have been conducted on simplified 2D models of micropumps. Furthermore, Ha *et al.* and Nisar *et al.* have presented analyses of piezoelectric micropumps using multiple code coupling

(MFX) capabilities in ANSYS and CFX simulation tools,[222,226] which require only one interface between the physical fields (fluid–solid). However, in electrostatically actuated micropump analysis for example, the design is more advanced as there are two interfaces between fluid–solid and solid–air fields.

Numerical modelling is a critical tool in the development of novel MEMS micropump models complemented with SAW-based wireless and secure interrogation and electrostatic actuation, targeting implantable drug delivery applications and host of other applications such as LoC, remote sensing, DNA sequencing and flow cytometry.[223,227,228]

For ease of understanding, a brief illustration of the operating concept of an example micropump is provided in Section 8.4. To the best of the authors' knowledge, for the first time, they have carried out a 3D multifield analysis of an electrostatically actuated valveless micropump using multi-field (MFX) capabilities in ANSYS, simultaneously combined with CFD capabilities in ANSYS–CFX for microfluidic analysis. This chapter therefore presents a detailed FEA-based CFD analysis, as well as a successful methodology to perform complex FSI analysis of MEMS micropumps.

8.4 Operating Principle of an Example Micropump

The implantable micropump is designed to achieve secure and wireless interrogation by eliminating any active electronic circuitry in the micropump. Figure 8.1 shows a concept drawing of a SAW device-based passive micropump. The valveless micropump structure is mounted on top of the the SAW device. In general, such a micropump structure consists of a thin conductive microdiaphragm, a fluid chamber and inlet and outlet diffusers. The gap between the conductive diaphragm and the SAW device is a few micrometers for effective low-power actuation.

The micropump operation is as follows. Based on the control signal, an electrostatic force is generated on the conductive microdiaphragm, which in turn creates a compulsive and repulsive force on the microdiaphragm. Since the conductive plate is a thin flexural plate, it vibrates as a function of the applied electrostatic field, enabling its use as an actuator for flow modulation.

8.5 Theoretical Analysis

In most of the microdiaphragm-based micropump mechanisms, the deflection of the microdiaphragm is very small compared to the typical length of the microdiaphragm.

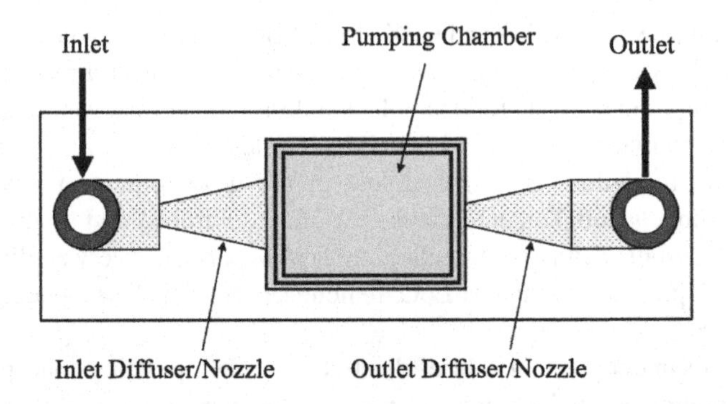

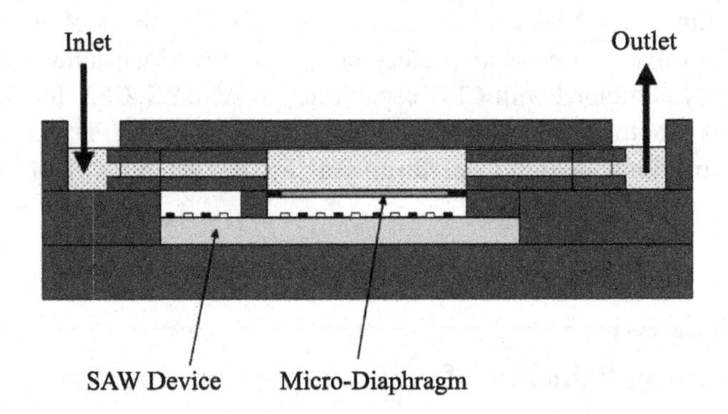

FIGURE 8.1
2D views of an example micropump structure. Top: Top view of the micropump; consisting of inlet and outlet diffuser elements for flow rectification, pumping chamber and microdiaphragm with a conductive coating. Bottom: Side view of the micropump; pumping chamber with the microdiaphragm is placed on top of the output IDT of the SAW device.

Therefore, the bending theory of plates is applicable and the transverse deflection for the microdiaphragm can be expressed as:[224, 229]

$$\mathbb{D}\nabla^4 W_D + \rho_D t_D \frac{\partial^2 W_D}{\partial t^2} = F - P, \tag{8.1}$$

where the bending stiffness $\mathbb{D} = \frac{\mathbb{E}t_D{}^3}{[12(1-\nu^2)]}$, $\mathbb{E}$ is the modulus of elasticity, t_D is the microdiaphragm thickness and ν is the Poisson ratio of the microdiaphragm material. W_D is the instantaneous deflection of the pump diaphragm, ρ_D is the density of the microdiaphragm material and ∇^4 is the two dimensional double Laplacian operator. Additionally, F is the actuating force acting on the microdiaphragm as shown later in Eq. (8.2), while P is the dynamic pressure imposed on the microdiaphragm by the fluid as shown in Eq. (8.3). The existence of F and P in this micropump structure is considered in detail in following sections.

8.5.1 Actuation Force

For a micropump mechanism based on, for example, electrostatic actuation, the electrostatic force applied on electrostatic plates can be described using the parallel plate capacitor effect,

$$F = \frac{1}{2}\frac{\varepsilon A \Phi^2}{(h - W_D)^2}, \tag{8.2}$$

where ε is the dielectric coefficient of the medium between the plates, A is the effective plate area, W_D is the instantaneous deflection of the microdiaphragm, h is the initial plate spacing and Φ is the applied electric potential between the plates.

However, in a SAW device-based electrostatic actuation, electrostatic force is generated due to the time-varying electric potential between the SAW device and the conductive microdiaphragm. This results in a more complex force than the force shown in Eq. (8.2).

8.5.2 Microfluidic Pressure Variation

The mass transport in microfluidic devices is generally dominated by viscous dissipation, while inertial effects are generally considered to be negligible.[221] In an initial concept model of a micropump, the fluid flow can be considered to be incompressible. Furthermore, non-slip boundary conditions (non-turbulent flow) can be assumed to exist at micropump walls. The governing equations for viscus, incompressible fluid flow can be written using Navier–Stokes equations and mass continuity equation as

$$\rho_L \frac{dV}{dt} = \rho_L g + \mu \nabla^2 V - \nabla P \quad \text{and} \tag{8.3}$$

$$\frac{\partial \rho_L}{\partial t} + (\nabla \cdot V)\rho_L = 0, \tag{8.4}$$

where V is the fluid velocity vector, μ is the viscosity, g is the gravitational acceleration is the density and P is the dynamic pressure of the fluid.

In a MEMS micropump model, these governing equations show that the actuation force, the deflection of the microdiaphragm and the flow of the working fluid are always coupled during the pumping process. In order to analytically determine the deflection of the microdiaphragm due to excitation force, Eqs. (8.1)–(8.4) are required to be solved simultaneously.

In general, the analytical modelling of full 3D fluid flow requires complex algorithms and extensive computational effort. Nevertheless, various researchers have successfully analysed such scenarios through numerical analysis and mathematical methods such as FEA-based CFD.[222, 230] As a novel contribution to the microfluidic device analysis, a multiple code coupling-based simulation methodology is deployed to analyse the complex coupling between multiple physics fields. Therefore, the actuation force field, structural and fluid field couplings are simulated and the FSI analysis of an example micropump is developed and presented in this chapter.

8.6 Model Development of an Example Micropump

A 3D model of an example micropump based on SAW actuation is presented in Fig. 8.2. Two interface boundaries are identified in the micropump model: (1) air–structure interface (ASI), which is between the microdiaphragm and the air gap, and (2) FSI, which is between the microdiaphragm and the fluidic chamber. Half symmetry is exploited to reduce the simulation time and CPU usage, without any compromise in accuracy.

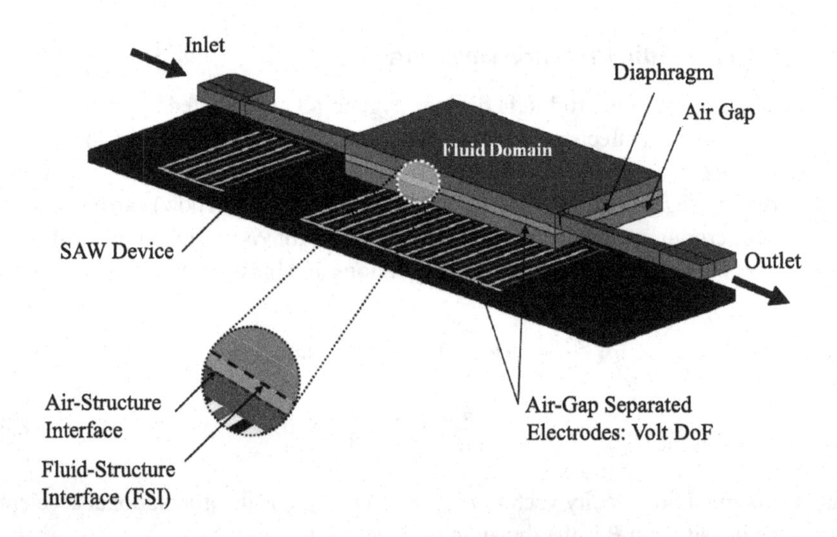

FIGURE 8.2
CAD model of the 3D valveless SAW based micropump example. The pumping structure is placed on top of the SAW device for multifield analysis, consisting of two interface boundaries. (1) ASI: between the microdiaphragm and the air gap and (2) FSI: between the microdiaphragm and the fluidic chamber.

8.7 Modelling and Simulation

Based on the example micropump structure, a full 3D analysis of the device is presented, without compromising any model feature. The CFD analysis was carried out in two stages — the initial design stage and the final performance analysis stage. During the initial design stage, firstly, the aforementioned simplified 3D model was developed and analysed. Thereafter, a full 3D model is developed and CFD

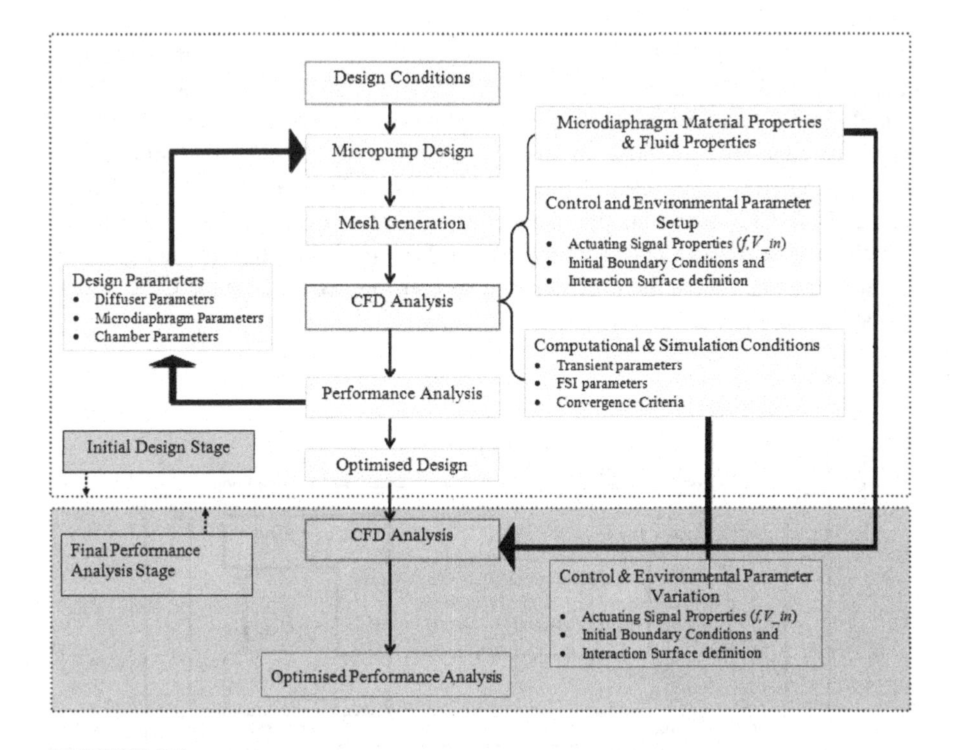

FIGURE 8.3

Design and performance analysis stages of FEA-based CFD analysis of MEMS-based 3D micropumps. The CFD analyses consist of two stages — the initial design stage and the final performance analysis stage. At the initial design stage, the performance of the micropump is investigated for various material properties, control parameters and environment parameters. For a specified set of such parameters, a repetitive analysis is carried out by varying design parameters until the performance is optimised. Once the design is optimised for its performance, various analyses are carried out to specify the final performance characteristics of the micropump.

analysis is carried out. During this analysis, the performance of the micropump can be investigated for various material properties, control parameters and environment parameters. Then, for a specified set of such parameters, as depicted in Fig. 8.3, repetitive analysis was carried out only by varying design parameters until the performance is optimised. Once the design is optimised for its performance, various analyses are then carried out to specify the final performance characteristics of the micropump.

In conducting FEA-based CFD analysis, it is important to set accurate convergence criteria and relaxation factors in the FSI analysis of microdevices.[36,231] Therefore, care is taken during the initial iterative analysis to fine tune such parameters for smoother simulations.

To solve the FSI problem, which consists of multiple interfaces, the multiple code coupling-based solution procedure of ANSYS and ANSYS–CFX can be utilised. A multifield solver with multiple code coupling is best suited for 3D FSI analysis due to its capability to carry out more physically complex and larger simulations with distributed physics fields. Figure 8.4 depicts the implementation of the ANSYS and ANSYS–CFX-based multifield coupling solver process for MEMS micropump modelling. The ANSYS code acts as the master and reads all MFX commands, does the mapping and serves the time step and stagger loop controls to the CFX, which acts as the slave.

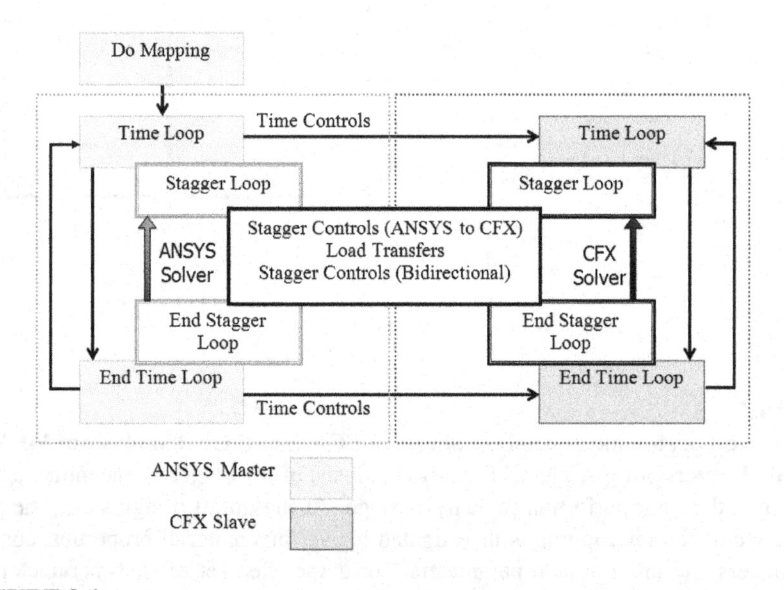

FIGURE 8.4
Multifield solution process for the CFD analysis. This demonstrates the interaction between various solvers during the FEA-based CFD analysis.

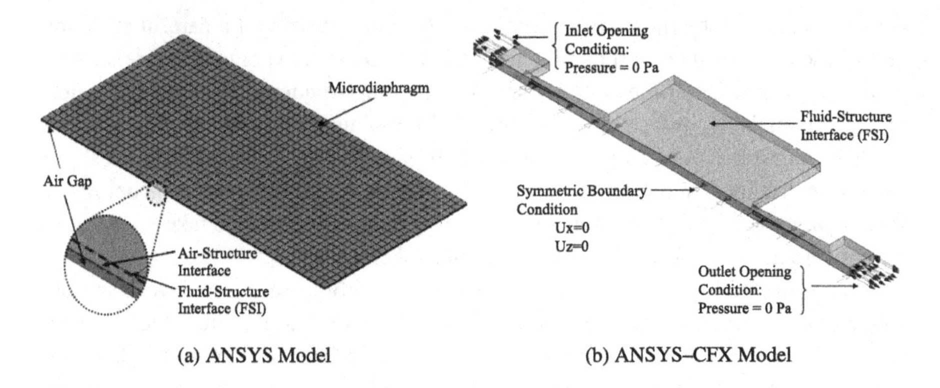

FIGURE 8.5
CAD models related to the structural and fluid region modelling. (a) ANSYS model of the example micropump consisting of the microdiaphragm and the air gap. Direct coupled field analysis is used to model the solid model consisting of the microdiaphragm and air gap, to determine the effect of forces in the electrostatically coupled microdiaphragm. The top and the bottom surfaces of the microdiaphragm interact with the fluid region and the air gap, respectively. (b) CFX model of the example micropump. Half symmetry is exploited, and the inlet and outlet pressure are set to 0 Pa to investigate the flow rectification effect.

In multifield analysis, the interaction at the ASI was simulated in ANSYS master code, using direct coupling analysis. The interaction at the FSI was simulated in both ANSYS master and ANSYS–CFX slave codes. During this FSI, forces get transferred from the fluid region to the solid structure and displacements get transferred from solid structure to the fluid region. The detailed setups of the 3D solid and fluid models are provided in Fig. 8.5. The air gap-coupled microdiaphragm is completely modelled in the ANSYS master code using a suitable element type for the microdiaphragm for the air gap. Similarly, a suitable element type was used to model the fluid domain, considering usability features of the element for the calculation of 2D and 3D velocity and pressure distributions in a single phase, Newtonian fluid.[232] During the CFD analysis, properties of water can be used to simulate the fluid flow for preliminary model development.

8.8 FEA-based CFD Simulations

Based on the design approach elaborated in the previous section, FEA-based CFD analysis is carried out for an example micropump structure. As the multifield problem is nonlinear in nature, the convergence is not guaranteed without a careful selection of multifield solution control parameters such as various convergence and relaxation parameters, multifield time controls and the coupling frequency between the fields. In such an analysis, instabilities could result from a number of factors; the solution matrices may have poor condition numbers because of the finite element mesh, or very large gradients in the actual solution. Moreover, the fluid phenomena being observed could be unstable in nature. Therefore, special care has to be taken to achieve the convergence of results during the flow analysis.

Therefore, multifield solver parameters were carefully set to achieve a smooth convergence of results. In FEA, factors such as the selection of incorrect element types, coarse mesh density and force convergence criteria are known to affect the stability of FEA results.[232] During the initial modelling of the micropump, element types were carefully selected to match with the solid and fluid fields, and the model was finely meshed using small element sizes. Thus, the remaining constraint, the convergence criteria was left for investigation. Especially, the force convergence criteria in the solid model (related to the direct coupled solver parameters), was found to be a critical parameter that has a direct influence on smooth and stable results. In order to demonstrate the influence of such parameters, results obtained through various force convergence settings are shown in Fig. 8.6. Based on the results, it is evident that

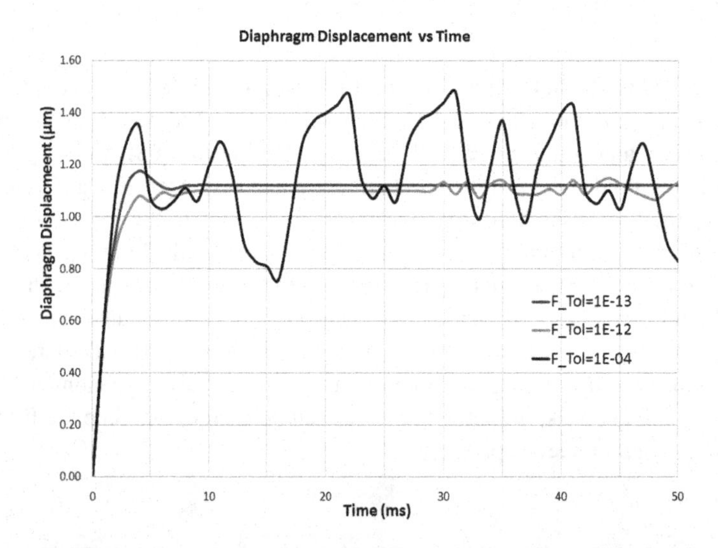

FIGURE 8.6
Force convergence analysis of the micropump. Variations in the dynamic behaviour of the example micropump for various force convergence parameters for a steady state analysis of the model.

the careful selection of convergence values during the preprocessing stage produced stable results.

Following a similar iterative method, CFD results were optimised for the remaining multifield parameters during 3D analysis. Then, the displacement and fluid flow characteristics and the damping effects were investigated. The results are presented in Fig. 8.7, where the damping effect caused by the fluid is clearly observed.

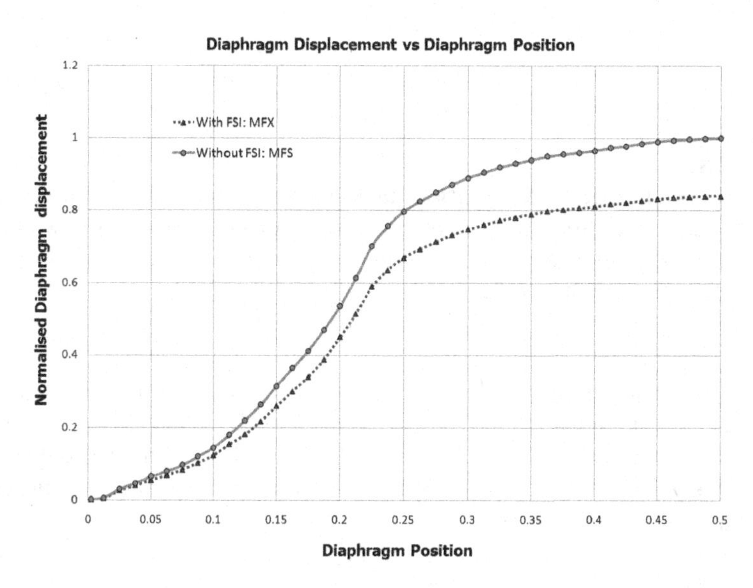

FIGURE 8.7
FEA-based deformation results of the microdiaphragm. Microdiaphragm deformation at the peak input signal, along half cross-section of the microdiaphragm, for simulations with and without FSI.

8.8.1 Inlet and Outlet Flow Characteristics

Once the initial setup parameters are decided based on preliminary analysis, a full 3D CFD analysis can then be performed. During the simulation, it can be noted that as the amplitude of the control signal is increased, the displacement of the microdiaphragm also increases and a higher chamber volume is created. Hence, the pressure inside the chamber is decreased compared to the pressure at the inlet and outlet. Further, as the amplitude of the control signal is decreased, the microdiaphragm displacement also decreases, and the relative chamber pressure is increased. This periodic variations in pressure difference cause the supply and pumping actions in the valveless micropump.

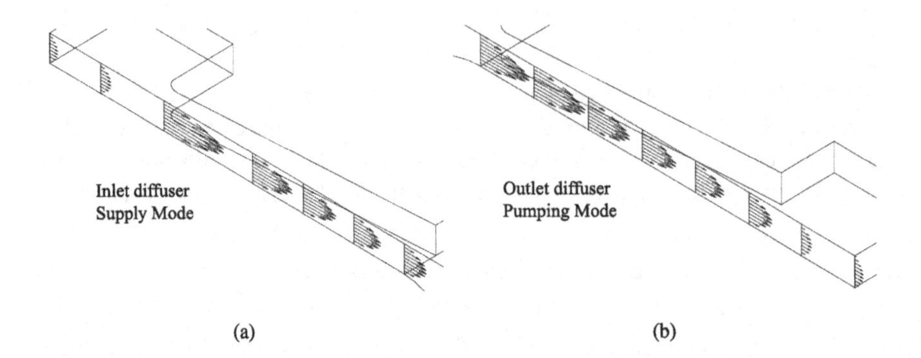

FIGURE 8.8
Inlet and outlet velocity distribution during supply mode and pumping mode of the example micropump. For any given step number, maximum flow velocity is observed at the center of the inlet and zero velocity is observed at diffuser walls, inline with the laminar flow behaviour.

Multifield analysis-based inlet and outlet flow patterns during the supply and pumping modes are presented in Fig. 8.8. Generic to all the plots, the laminar flow profile is observed across the inlet and outlet diffusers. Here, discretised flow velocities at diffuser walls are zero, and the velocity gradually increases towards the centre of the diffusers.

8.8.2 Flow Rate Estimation

In such flow analysis that is centred on FEA-based CFD simulations, the volume-averaged flow velocity at the inlet and the outlet can be computed and compared to evaluate the overall flow rate of the micropump, during the postprocessing stage of the model. The volume-averaged velocity $\overline{v}$ is defined as:[233]

$$\overline{v} = \frac{1}{A}\int_A u_a \, dA, \tag{8.5}$$

where u_a is the axial velocity across a cross-section of the diffuser and A is the cross-sectional area of the inlet and outlet diffuser necks. Volume-averaged flow velocities for inlet and outlet are plotted against simulated time steps in Fig. 8.9. During the supply mode, the volume-averaged flow rates at the inlet are always higher than that at the outlet. In contrast, during the pumping mode, the volume-averaged flow rates at the outlet are higher than the inlet. A positive rate indicates that fluid is flowing in the chamber, and a negative rate indicates an outward fluid flow from the chamber.

These results clearly demonstrate the pumping effect of the SAW device-based valveless MEMS micropump. The simulation-based average flow rate (Q_{avg}), of the micropump can be calculated as

$$Q_{avg} = \frac{2}{T} \int_0^{T/4} Q(t)\, dt, \tag{8.6}$$

where T is the time period of the control signal and $Q(t)$ is the instantaneous volume averaged flow rate, which can be computed as $Q(t) = A\overline{v}$.

Based on Eqs. (8.5) and (8.6), and CFD results of the micropump, the average flow rate of the designed micropump can be calculated. For the example micropump presented, the flow rate is calculated to be 43 μL/min. Such flow rates under low-powered actuation indicates competitive performance characteristics, compared to the other reported analysis on valveless micropumps.[36,234] In addition to this point, the computational modelling of such micropumps and the performance estimation techniques can be utilised to evaluate the power requirement for these devices, and compared with the other reported valveless micropump models in the literature for improvements of the device at the design stage as a low cost methodology over fabrication.[36,38,220,221,234]

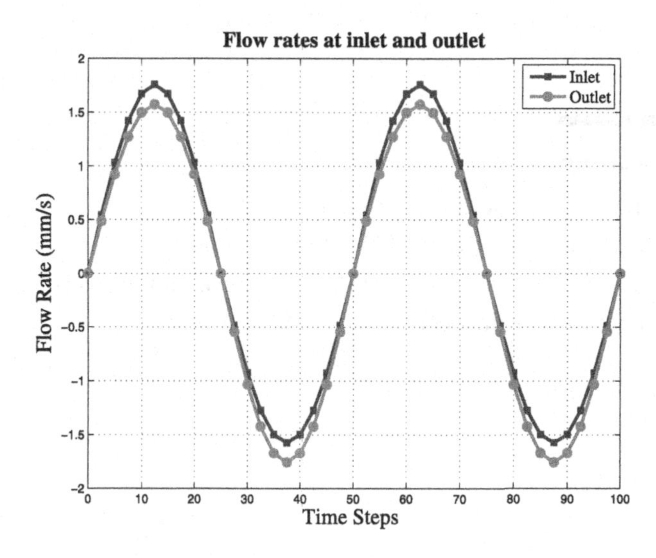

FIGURE 8.9

Averaged flow velocities at inlet and outlet. During the supply mode, the volume-averaged flow rates at the inlet are always higher than that at the outlet. During the pumping mode, the volume-averaged flow rates at the outlet are higher than the inlet.

8.9 Summary

In this chapter, the advanced computational modelling and analysis of MEMS micropumps for biomedical applications were discussed. A novel 3D modelling and simulation methodology was presented, which is a combination of 3D multifield analysis and multiple code coupling capabilities in commercially available modelling and simulation tools. More importantly, the FSIs between multiple physics fields have been effectively modelled and investigated for detailed 3D models of MEMS micropumps. A combination of FEA and CFD numerical methods are utilised for the complex interaction analysis of mircofluidic devices with multiple physics field interactions such as fluid, solid and electric.

The utilisation of advanced modelling and simulation procedures for the accurate evaluation of microfluidic devices, as well as the capability of the system to identify and quantify flow behaviour can result in clinically safe designs and provide medical experts with a leading edge in saving patients. Importantly, this interaction analysis methodology can be successfully utilised as a platform for computational modelling, analysis, optimisation and verification of a range of microfluidic devices and systems. For example, acoustically driven micromixers and passive sensors, LoC devices and implantable total micro-drug delivery systems are some of the systems that can be simulated and analysed by extending the method followed in this work.

8.10 Questions

1) List the two main causes for hypertension.
2) Name three commonly used anti-hypertensive drugs.
3) What are the factors that affect an electrostatic force between two charged plates?
4) What are the three main components of a SAW device?
5) What is the advantage of using a SAW device combined with a microdiaphragm when designing a micropump?

9

Engineering and Production Management for Biomedical Devices

CONTENTS

9.1 Overview of Product Development Strategy 127
  9.1.1 Product Conceptualisation 129
  9.1.2 Market Survey and Strategic Alliances 129
  9.1.3 Design, Prototyping and Product Development 130
  9.1.4 Testing and Commissioning 131
  9.1.5 Technology Protection 132
  9.1.6 Marketing and Sales 133
9.2 Product Development Management System 134
  9.2.1 Systematic Management 134
  9.2.2 Standard Operation Procedures 134
  9.2.3 Material Resource Planning 135
9.3 Organisation Charts 135
9.4 Project Management 139
  9.4.1 Fundamentals of Project Management 140
  9.4.2 Project Life Cycle 142
  9.4.3 Project Management Team 143
9.5 Summary 146
9.6 Questions 147

9.1 Overview of Product Development Strategy

This section focuses on the scientific management of design, engineering and production that pertains to the biomedical device. The strategising of organisation structure, procedures of projects implementation, as well as quality assurance is crucial to the entire development of a biomedical device product. A scientific management of projects within a biomedical device manufacturing company is the *sine qua non* of success in efficient and effective product development. The value of well-established scientific management is determined by this effectiveness and efficiency that it can bring to its research and development facility, manufacturing plant and all other departments integrated within the organisation.

For scientific management, the general decision making procedures are: (1) information retrieval and data sorting, (2) analysis of the situation and potential obstacles, (3) generation of approaches and (4) strategisation of solutions and contingency plans. Once an organisation becomes successful and self-substainable, it presents an opportunity to expand manufacturing facilities to increase its competitive edge. In traditional manufacturing, the most successful business model is to have products that can be released in sufficient quantities into the market in order to sustain expansion and counteract competition. However, during highly specialised product development such as biomedical devices, the shift from quantity to quality is important as, in some cases, a patient-specific design will be more medically viable as compared to various standard design products ready on the shelf. Therefore, such highly specialised products impact on the reputation of the organisation and the market share that it can potentially capture, which is linked to profits and company expansion. Biomedical products used for medical applications generally involve specialised technical knowledge that is not widely available among merchandising agencies and distributors.

From the organisational point of view, the biomedical device manufacturer needs to have the basic research, product engineering, manufacturing facilities and a commercialisation unit. The research and development department designs the biomedical device based on the required specifications and clinical needs. The industrial department is responsible for procedures that improve the efficiency of device procurement. The procurement department retrieves the blueprint from the product design department, together with specifications of the product route during manufacturing, as well as the design tools, rigs and fixtures used during the production. The manufacturing department mass produces the biomedical device. Each of these departments plays an important role in the organisation and are often integrated together in an essential framework, which propels the need for a good management strategy.

A biomedical device production system arises when procedures have crystallised into standards to coordinate and integrate the product research and development with manufacturing. In this chapter, we examine the principles adopted in manufacturing biomedical devices. In particular, flow charts and diagrams depicting the structure for strategising for developing and manufacturing management such a product are prepared.

Based on Chapter 1, the first step in biomedical device development is the conceptualisation and research in the feasibility of the operation. Obtaining a full grasp of the design concept and an understanding of its feasibility requires an analysis of its functional needs and requirements. Having a good prediction of patient suitability involves an extensive market survey. Knowing if the device has attained safety standards requires testing and commissioning. Other procedures that ensure the sustainability of product development and manufacturing as well as the survivability of the organisation that deals with this product, such as patent application, medical insurance, marketing and sales, are required. In addition, an efficient material flow leading to high production and well-maintained quality assurance should be implemented. We will look at each of these important development procedures or operations. The duration for product conceptualisation leading to its eventual production and marketing depends on the complexity of the medical product and the extent of invasiveness to humans.

9.1.1 Product Conceptualisation

Conceptualisation of a product design enables the achievement of novel or effective product engineering. Design conceptualisation examines the following questions: the type of product that the patients most require, the market potential for this product, the ease of clinical usage or surgical implantation, as well as future technological advancements that would enhance this product. In general, the conceptualisation process is usually supported by functional needs and requirement analysis, technological survey and market analysis. The functional needs of a device will determine the operation mechanism as well as the research methodologies used in developing it. The product research and development leading to its manufacture is a long process requiring scientific and robust standard operation procedures and guidelines for engineering production. The visualisation of a design gives rise to the conceptualisation of the future product usually with the aid of CAD, or a modelling platform of the variant products, enabling the implementation of their operation simulation. Product visualisation can guide designers to improve design concepts and achieve realisation of the optimal design. It becomes a tool for strategic design generation, and can be utilised at the marketing stage as well.

9.1.2 Market Survey and Strategic Alliances

The key purpose of a market survey is to determine the type of product that best suits the patients. This may be conducted in a hospital, whereby opinions from medical workers and patients may be of value to the survey. Financial exhausion is high when it comes to the development of new products and pentration of new markets. In this case, the engineering capability, the type of skills and technologies that are available or off-limits to the organisation should be informed. Competitive benchmarking can systematically calibrate performance against external targets, which can be a competitor or a partner, and determine the advantage of competitive collaboration.[235] If it is advantageous to collaborate, an alliance of engineering organisations and manufacturers can enable the improvement of design effectiveness, production efficiency and quality, which is why a thorough survey is sometimes needed to assess the feasibility of such an approach. In addition, such an action constitutes low cost strategies to building new process capabilities. A thorough market analysis will inject confidence to the organisation in its strategic plans and operations.

On the product development front, as engineers explore new research advances in the market and obtain feedback from the potential users of the biomedical device, product designs and technologies can be improved over time. To commercialise a new product into the market, the immense quantity of information from the industry, users and researchers should be properly managed and analysed. Failure to meet cost targets due to an overestimation of sales or investments can lead to a substantial loss of profits for the organisation.

9.1.3 Design, Prototyping and Product Development

As explained in Chapter 3, the design and prototyping of the researched biomedical device is crucial in determining the success of its mechanism and its safety operation range. A new biomedical device may involve a large-scale development program. Based on the many components of the development, tasks are assigned to each engineering department and these departments may comprise several sub-divisions. To ensure the satisfactory performance of the biomedical device, mechanical working parts should be standardised, and standard operating procedures should be given flexibility for minor improvements that would reduce costs and improve efficiency. The components of the product can be put together into a blueprint to provide a general product drawing. A CAD allows changes to be made such that it meets the safety threshold or achieves consistency with the approved design. The protoype enables the engineers to evaluate their designs. Typically, multiple re-designs are required prior to its finalisation.

As a technological product slowly progresses into the more advanced stages of its development with the incorporation of new technological components, an increase in its level of engineering quality, clinical/surgical safety and operational reliability can be achieved as demonstrated by Fig. 9.1.

The use of progressively improving research and engineering knowledge to devise medical-based solutions results in the design and development of biomedical devices in stages. The advances in its design are heavily pegged to technology developments and may take place over decades. This comes with the incremental incorporation of the formulated scientific principles into their devices. Let us briefly review the development of the artificial cardiac pacemaker, which is an electrical device that keeps the heart pumping in its normal and regular beating mode. The initial model was developed as an external support device that transmits pulses of electricity to the heart muscles via electrodes leads on the chest. The electrodes are delivered through the chest directly and contacting the heart. This enables the stimulation pulses to pass through the body and to the electrodes attached to the myocardium. However, the recipients of the device suffered infection at the entrance of the electrodes through

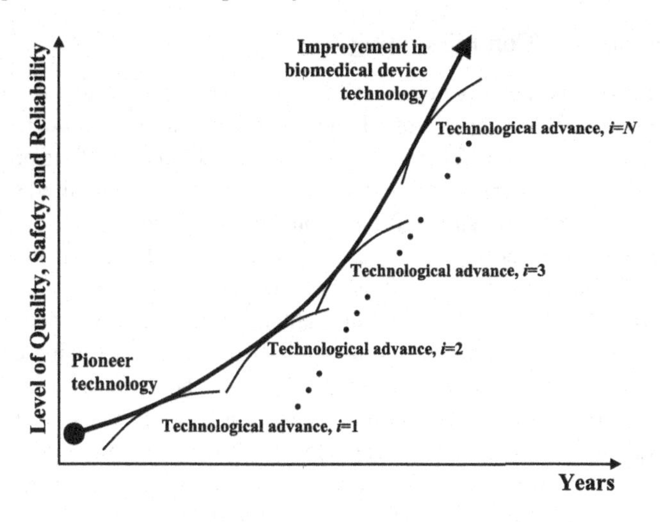

FIGURE 9.1
Technological advances in a biomedical device product. The improvement in the biomedical product quality, safety and reliability is influenced by the technological advances number $i = 1$ to N, and reflects the symbiosis between the technological evolution of the product and these individual advances over time. (Figure adapted with permission from Lienhard JH. *How Invention Begins: Echoes of Old Voices in the Rise of New Machines.* Oxford University Press, New York, 2008.)

the chest into the heart. Subsequently, researchers clinically trialled the use of the world's first fully implantable internal pacemaker that has its electrodes attached to the myocardium by thoracotomy. Unfortunately, solving one problem leads to another. Due to its infancy in implementation, the fine wires of the device experienced a high rate of failure, which might have led to the replacement of the entire device. Eventually, the most successful pacemaker emerged with the invention of a novel wire comprising an inter-wound helical coil of wire alloys that are in turn wound into a spring. Future technological developments led to the invention of the isotope-powered pacemaker, also known as the nuclear pacemaker, which gives rise to a power source that can last for the entire lifespan of the patient. This pacemaker is, by far, the most reliable and safe working product.

In the face of global competition, the research and development of new products are a necessity. Sometimes, the realisation of new design may require changes in processes related to its manufacturing. Due to a shorter product lifetime and dedicated development and production at its initial phase of research, costs are usually high. As such, a strong research and development capability that is paired with flexible assembly lines and manufacturing processes can give a competitive edge to organisations in churning out innovative products.

9.1.4 Testing and Commissioning

The frequency of biomedical device failure has increased due to the tremendous growth in medical technologies and a large number of biomedical devices entering the market in recent decades. Therefore, a thorough examination of the product prior to its entrance into the market is necessary. Testing to meet the required standards is carried out during the final stage of development, and is necessary for mechanism reliability and safety. Let us use the case of the heart valve as an illustration of the testing of biomedical device. The shortcomings in the existing modelling and testing of prosthetic heart valves are highlighted by the recently reported cases of post-market failure consisting of thromboembolism, in leaflet escape due to strut fracture.[236] Strict regulatory *in-vitro* evaluation must be performed on prosthetic heart valves, prior to passing the clinical standards before release to the market. This work has significant benefits as it puts together useful information for designing procedures and instruments used for the accurate and effective determination of the physiological flow behaviour of blood through the aortic valve, and such information will be utilised for reliable heart valve design, prototyping and testing.

High clinical standards and standardised testing procedures are important for heart valve testing as ineffective valves can endanger lives in the event of failure after a period of time. The Medtronic Parallel valve is a classic example of post-market failure due to thromboembolism, which resulted in patient fatalities.[236] Therefore, before the implantation of heart valves, the mechanism of the valve must be checked for operational safety, based on international standards such as the ISO 5840, CEN and FDA guidances. The European Community (CE) medical use approval, defined in the CE medical devices directive developed in 1993, is certified by conformity CE marking or the "CE mark". The FDA guidance document for the submission of medical use approval requests was issued in 1993, and revised in 1994 by the Division of Cardiovascular, Respiratory and Neurological Devices.[237] Particular requirements for cardiac and vascular implants, outlined by the British Standard BS EN 12006-1:1999, exist for the pulsatile flow testing of heart valve substitutes. The testing process should be able to assess the inherent risks in the product and the difficulty in rectifying it, and may even recommend the termination of its production if no remedy can be offered. The biomedical device may pass all tests, but manufacturers will normally ensure insurance protection of the device by applying medical indemnity for collateral damages.

9.1.5 Technology Protection

The product and process designs that consitute inventiveness sometimes result in patents based on a collaboration of work among the research and development team members in an enterprise. Due to the immense cost of performing research and development, organisations developing the state-of-the-art biomedical products view patent protection as a way of securing their research and development brainchild and prevent industrial competitors from entering the market. This is because their competitors can exercise the cheap option of dismantling the technology to discover the novel ideas associated with the product and then reverse-engineer them to generate the same, or if not, an even better technological product. As such, patents are usually an appropriate method to allow control over the inventive technology with a limited-time exclusive right, and to enhance the organisation's competitiveness in the biomedical device market. In that sense, a patent can be viewed as a *monopoly right* for the invention. Some of the large-scale organisations have their own patent specialists that are specifically responsible for patenting the research outputs by their engineers.

While patents are designed to enhance inventions and innovations, other schools of thought state that they form barriers to perform innovation.[238,239] This is because patents are cumulative, owing to the fact that a particular technological product may be subsequently evolved to an upgraded version and embody a further collection of inventions, resulting in parallel technological innovations to meet the changing market needs. However, patents restrict the exclusive rights to develop the technology for the other parties and starve off a chain of inventions that can potentially arise from their research.

A trade secret protection forms another alternative for securing the technological innovation or inventiveness, and can also ensure that the product is secure enough to be commercialised by the biomedical device manufacturer. The appropriation of trade secrets can be prevented by security measures such as imposing physical security and enforcement. This can be further achieved by stricter regulations on checking employees for possessing sensitive information such as product blueprints in hardcopies and in data storage devices, raw materials, by-products, customised equipment or experimental prototypes when leaving the organisation. Other measures include screening contacts in the form of electronic mails or telephone calls made to suspicious or foreign competitors in similar industries. In contrast to patents, trade secrets restrict the disclosure of an organisations technological innovations and inventions to the general public. In this aspect, some companies may demand that their employees maintain strict confidentiality of the secrets even after they are no longer contracted with them. For the more extreme cases, legal agreements may also be enforced such that the ex-employees are not allowed to engage any work contracts from their direct competitors. However, some organisations may relax this restriction by lifting that constraint until a specific waiting period is over.

9.1.6 Marketing and Sales

In general, the quality and performance of a company product determine its competitiveness in the market. In addition, by understanding the patient condition and selling the right product at the right time, higher sales can be achieved. Having a thorough understanding of the market enables the organisation to effectively strengthen its brand of biomedical device and the company itself. The use of marketing and sales skills is important here in this regard. New startups or partially mature companies trying to enter the market tend to ride on the wave of existing technologies, and adapt readily to the biomedical device market demand. The proximity of organisations in the global scale context increases with the advancement of technology and globalisation. The linkage of various companies in their supply chains may put a spread of its sales over a larger region of distribution. To meet competition from existing products, the organisation must identify the high-sales products. Research engineers constantly analyse their competitors' products and propose a flow of new products into the market. The technical data used by the sales department may be interpreted and supported by the research department at best, so that the sales effort can be improved.

9.2 Product Development Management System

9.2.1 Systematic Management

In a systematic design, engineering and manufacturing environment, the execution of procedures and orders undertaken throughout the development of a biomedical product by an effective system are desirable. Systematic manufacturing reflects the full development of a product within an organisation without the corresponding development of time-wasting processes. Such scientific management render executives or engineers to be more methodical. During the product design phase, the engineer at every stage of the process should receive condensed and summarised design blueprints or procedures that should have been carefully assessed at every stage of the analysis. In terms of production, manufacturing should be properly managed by a set of standard operation procedures and effective material resource planning, which we will introduce subsequently. Foresight and systematic planning can reduce risks and improve efficiency. Poor management, on the other hand, increases the complexity of design and production, which can (negatively) affect the quality of the product.

9.2.2 Standard Operation Procedures

Standard operation procedures define the scope of engineering and production. This information describes the methodologies for design, engineering and manufacturing, which affect the whole production operation. The establishment of rules facilitates the delegation of authority and responsibility to the relevant teams or departments.

Written design regulations can be as general as possible or detailed to the extent that they provide the exact operations of every engineering or manufacturing task in the system. These operation procedures should be flexible and versatile enough so that they are able to be evolved quickly as design constraints are modified based on different case conditions.

9.2.3 Material Resource Planning

Effective material flow pertaining to the manufacturing process is crucial. The managerial of the material resource planning system is the crux of the success of manufacturing. For example, the purchase of raw material and a material resource enterprise system requires effective management in order to allow the flow of raw materials and manufacturing into the production system. The logistics of production and equipment maintenance is required to ensure a smooth manufacturing hierarchy.

9.3 Organisation Charts

An organisation chart is a graphical representation of the relationships between various departments within an enterprise with a well-defined structure (Fig. 9.2). An organisation structure may be complex, often complicated by the problems of coordinating various departments. Although the growth of an organisation over time may result in the addition of new departments with more branches, it results in a more intricate hierachical structure, and often comes at the expense of efficiency in coordinating operations and difficulty in management.

The chief executive officer (CEO) is the head of the company and usually oversees the operations of the engineering, finance, sales, manufacturing and procurement departments. Under each department is a departmental chief who reports directly to the CEO. Let us limit our scope of interest by examining only the departments that relates to the design/engineering/production of the biomedical device. The engineering department is often closely related to the manufacturing department. The chief engineer is typically a crucial member of the manufacturing division and usually controls the product design and manufacture. The chief manufacturing officer's responsibility is to ease the flow of production. The chief inspector oversees and maintains the product quality. The superintendent of every department deals with particular functions of the manufacturing. The production manager deals with operations that assist the other production departments. The operations manager controls all operations related to planning and strategising of product development. The standards and methods department investigates and give orders for work plan tasks and the directing of the engineers, and outlines check details of the manufacturing.

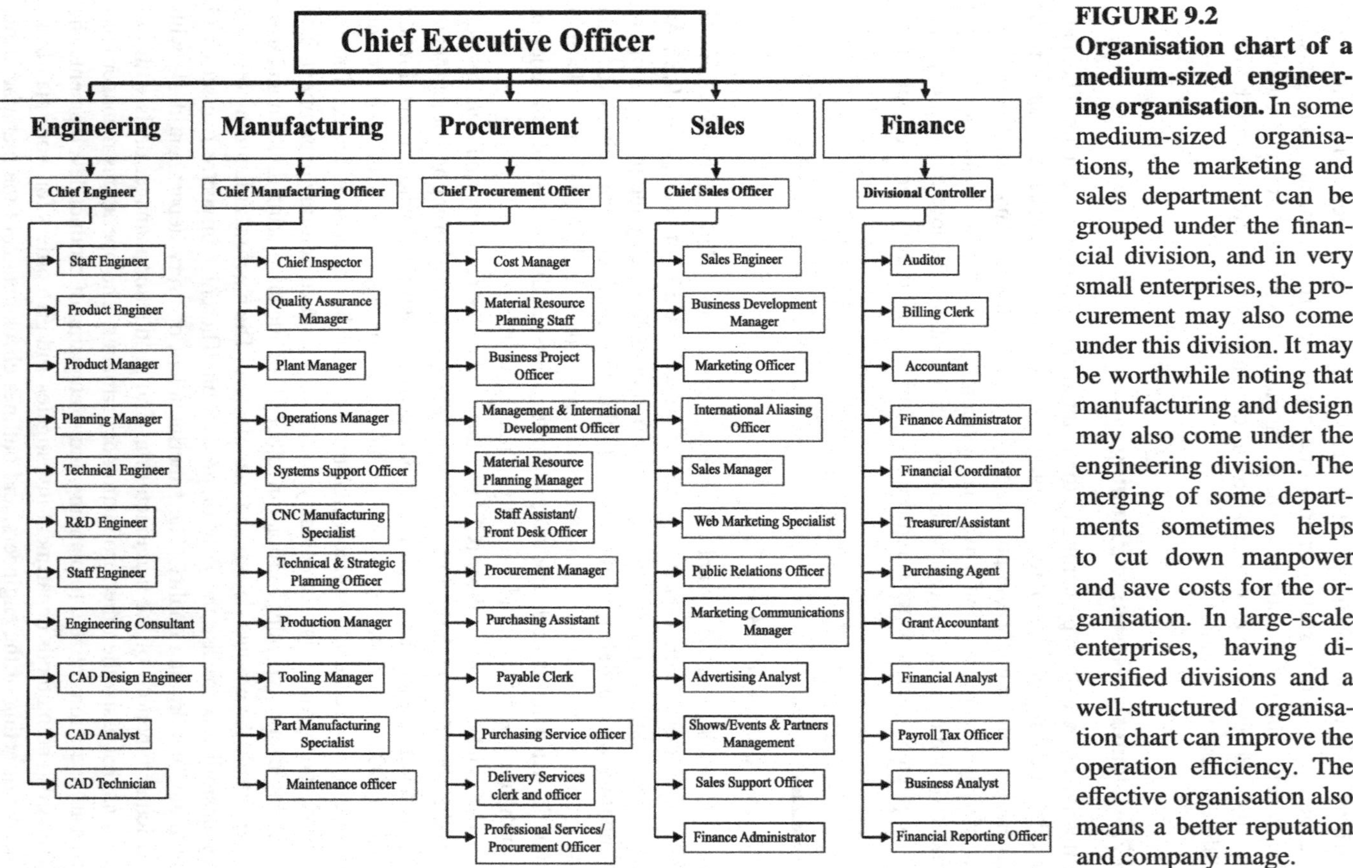

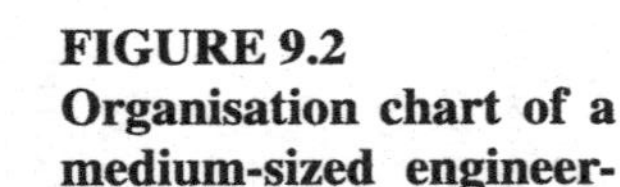

FIGURE 9.2
Organisation chart of a medium-sized engineering organisation. In some medium-sized organisations, the marketing and sales department can be grouped under the financial division, and in very small enterprises, the procurement may also come under this division. It may be worthwhile noting that manufacturing and design may also come under the engineering division. The merging of some departments sometimes helps to cut down manpower and save costs for the organisation. In large-scale enterprises, having diversified divisions and a well-structured organisation chart can improve the operation efficiency. The effective organisation also means a better reputation and company image.

The design team (comprising the CAD analyst and CAD technician) maintains the upgrading and innovation of designs based on feedback from the engineering departments in the organisation. In a larger organisation with a research team, the research and development department is responsible for the product innovation and generation and delivery of new product concepts.

The coordination by the design/engineering/production department facilitates and synchronises the planning, organisation and operation of stages from product conceptualisation right down to manufacturing and production, and insure that the time cycle for production is minimised. The issues and demands during production that are handled by the coordination process vary with the type of organisational structure implemented for the aforementioned departments, the operating procedures and policies established within the department, the capabilities of the engineers and the complexity of the biomedical devices.

Furthermore, in most cases, there exists a need to launch new products to meet the existing market round the clock to maintain competitiveness. As such, sometimes to form a committee that serves a specific purpose such as the development of a biomedical device product, a group of dedicated specialists are employed and called the new product development advisory board committee. The development of novel products and their systematic innovations require information sharing, generation of new ideas and coordinated upgrades throughout the entire design realisation process. Let us examine some of the key components in this committee that constitute the organisation's internal management.

A new product development advisory board committee consists of the board of directors, who are the chiefs in their respective departments (engineering, manufacturing and sales) heading this committee and presiding over the committee's decisions and actions. A level down, the engineering/production team (product manager, production manager and plant manager), the manufacturing team (tooling manager and maintenance officer) and the cost analysis team (sales manager, procurement manager and cost manager), as shown in Fig. 9.3, work under the directors' instructions. The production manager is involved in the planning and organisation of production schedules for new designs. The product manager and the plant manager take charge of production line updates for preliminary product design manufacturing. Important decisions such as the plans of changing products or product designs, as well as new items for marketing and the involvement of sales and production related to the product are evaluated by the sales manager, the cost manager and the procurement manager that form the cost analysis team. The budgeting of research and development costs is implemented to control expenses within a certain amount and in a balanced relationship among the various divisions. This would be the key responsibility of the cost analysis team. Budgeting that exceeds a pre-defined limit typically needs to be approved by the board of directors. The sales manager need not sit in on the committee meetings unless sales matters are involved. The tooling manager from the tool department upgrades the manufacturing tools to suit new products and maintain their working conditions. Working with this department, the maintenance officer looks after the new machinery and manufacturing equipment, and together, they form the manufacturing team. The chief inspector from the quality assurance department ascertains the quality of product outcomes within a desired safety threshold, and ensures the safety of usage by patients. The committee also convenes to discuss technological progresses and evaluate changes in existing designs, or any new ideas that may be implemented due to technological advancements or breakthroughs.

The reduction of costs in a manufacturing system is also a key concern faced by any organisation. That is why for most organisations, in conjuction with the new product development advisory board committee, a quality assurance team may be implemented to streamline manufacturing processes and reduce redundant costs. This team can also operate independent of the new product development advisory

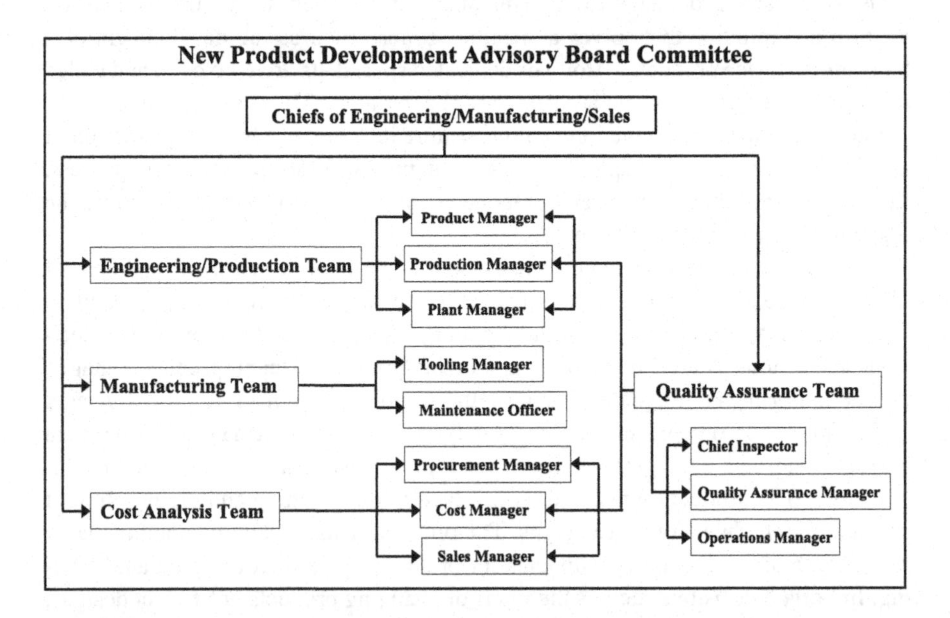

FIGURE 9.3
Structure of the new product development advisory board committee. In an effective organisation, an advisory board committee is required to oversee the research and production of specific biomedical devices such that their configurations are properly maintained, and that the processes from product design and development, right down to manufacturing are all up-to-date. An appropriate structure of the committee comprises a few key individuals (usually formed from other departments) playing relevant roles in the biomedical device development.

board committee during normal production. Typically and finally, a coordination of routine operation, status of orders, causes of excess stocks and progress of manufacturing programs is needed as well. The quality assurance manager, cost manager, the operations manager and the managers in the engineering/production team, the manufacturing team, as well as the cost analysis team in this committee work together effectively to enable the delivery of the biomedical device product at the lowest cost and highest quality. Lean manufacturing is a concept that can enhance the organisation operational standards and its efficency, which can be the product design, supplier management, information analysis, etc. It develops mechanisms in improving all structures, processes and skills within the organisation. Related to the quality assurance perspective is the Six Sigma methodological process that delivers products and services with quality up to a 3.4 defects per million parts. This philosophy, which was initiated by Motorola in 1986, has widely been implemented in organisations manufacturing a large quantity of products. It constitutes a statistical term that measures the quality of a process or manufactured product. The main task of the quality assurance department in a manufacturing organisation is the systematic reduction of defective products up to the Six Sigma level.

In summary, the product development advisory board committee and Six Sigma quality assurance team pertain to cross-functional teams (CFTs). Such CFTs can be useful for getting the managers to see beyond their functional or regional scope of responsibilities. The CFT is such that it typically comprises the company's middle management and is limited in size. It usually has a low level of hierachy to enable effective communication and to address the scope of the problem efficiently.

9.4 Project Management

Project management, which came to recognition as a scientific concept in the 1950s, is a useful tool that is implemented in organisations that pertain to a variety of disciplines such as product engineering, research and development, manufacturing, building construction, etc. Standards in PM have been proposed, and one example would be *A Guide to the Project Management Body of Knowledge*, which is abbreviated as the PMBOK Guide, is first published by Project Management Institute (PMI) in 1983 as the American National Standard. It describes the generally accepted and typical project management style, guidelines and practices that are common to "most projects most of the time". Project management is an integrated process or methodology for portfolio, program and process management leading to deliverables such as a biomedical device. Effective project management can speed up a company's operational efficiency.

9.4.1 Fundamentals of Project Management

In general, a project comprises at least four stages as shown in Fig. 9.4. Initiating a biomedical device research and development (R&D) project is implemented as the first stage of the project management (Stage 1). Prior to executing any concrete action, adequate planning should be carried out to obtain the approval of the board of directors for project initiation (Stage 2). This does not pertain to a simple announcement or launching action. It requires an enormous amount of preliminary work that mainly consists of idea generation, the review of present operations and the analysis of scope, time, budget, risk, resources, etc. to achieve the required quality goal. Quality is one of the major components of product scope whereas time, cost and product quality are the three main determinants for the overall quality of this project performance. Executing, which mainly involves project controlling and monitoring, ensures the success of the project and puts the blueprint of the biomedical device and its R&D plan into practice (Stage 3). Not all projects pertaining to different biomedical devices should involve all the four stages. In fact, some projects may go through planning and executing repeatedly in a loop (Stage 3.1). Monitoring and controlling of the project are performed by the project manager, and sometimes taken over by its team leaders (Stage 3.2). This involves ensuring that the project is on schedule and within its financial budget. Finally, closing and evaluation comes when the product is fully engineered (Stage 4). It is sent for in-house mechanical testing or clinical testing in the laboratories. Upon successful evaluation of its effectiveness and efficiency, it will be ready for commissioning.

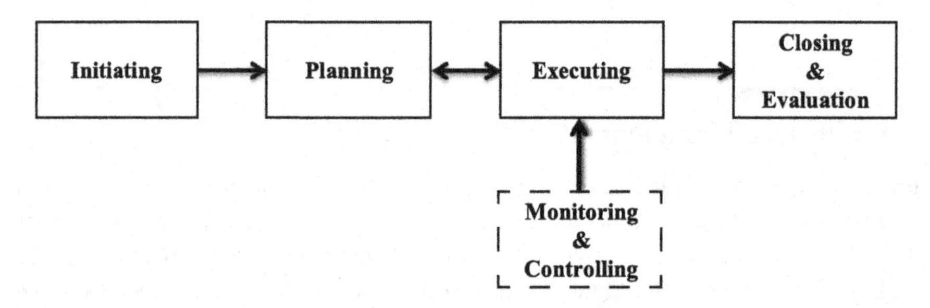

FIGURE 9.4
The four phases of a typical project management. An effective project management flow is made up of four stages, namely, the *Initiating*, *Planning*, *Executing* stages that are concurrent with the *Monitoring and Controlling*, and finally *Closing and Evaluation* stages.

The six major constraints of a project management form a close relationship with each other (Fig. 9.5). Any change in one factor will have a certain amount of impact on the others. To generate a creative idea, communication with main users, i.e., patients and medical practitioners, are needed to record ergonometric and clinical requirements. The next stage is to collect and evaluate engineering solutions and concepts, and the final decision making lies with the biomedical device R&D project manager.

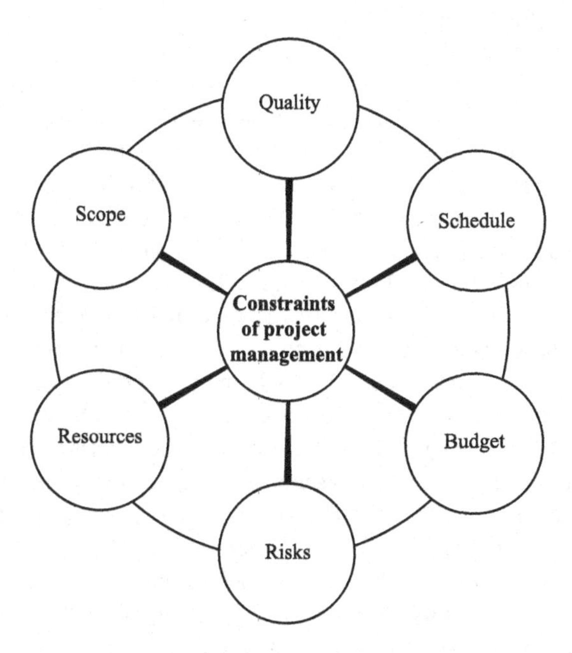

FIGURE 9.5
Six major constraints faced during project management. Projects are constrained by the following aspects: Quality, Schedule, Budget, Scope, Resources and Risks. Each of these constraints has a role in influencing the success of the project and should be well examined prior to and during the running of the project.

For project management in the field of biomedical device R&D, the terminology scope involves requirements in two aspects — the project scope and the biomedical product scope. The former pertains to the work that is required in order to deliver the biomedical device that meets the specified features and functions. In contrast, the latter is concerned with the features and functions that characterise the biomedical product. Scope management needs to be in place in order to keep the project from entering into a scope creep. This usually occurs when extra requirements are not included in the original plan. For example, the modification of an engineering aspect of the biomedical device that conflicts with safety in clinical usage may suddenly arise and be pushed forward, resulting in project failure to adjust the schedule and budget in time. These constitute risk to the project. Detailed documents stating contingency plans should be prepared during the planning stage, which can be used to monitor and control what is within and beyond the initial plan in its execution stage. It is also worthwhile noting that the scope analysis should cover the market and competition analysis, which are indispensible for positioning the biomedical device as a reliable product in the medical industry.

Project schedule planning is important in order to estimate the duration that will be required to complete and deliver the product, as failure to meet the strict deadlines may result in overruns of cost. As such, it is critical to outline each stage of the schedule based on milestones. Cost analysis influences the finances of the project and in particular, the budget for the biomedical device development and the profits it can bring in to self-sustain the organisation that manufactures this product.

A formal and detailed plan should be then drafted on the basis of the information collected and analysed during the first stage. Although plan building is a complex and painstaking process, it is worthy of attention and efforts as it establishes the foundation for the success of the R&D as well as the production of the device. To achieve a concrete plan, it is required to state the (1) technical goal: what does the biomedical device achieve?, (2) work breakdown structure: how to achieve a successful R&D of the product and commission it?, (3) personnel arrangement: to which group of engineers, i.e., from which disciplines, to assign to the specific engineering tasks?, (4) schedule: when to kick off and terminate each development phase? and (5) budget: how much would each phase of the whole project cost? As mentioned earlier, risks are likely to arise due to unforeseen circumstances and contingencies should also be included in the initial plan. In summary, during the planning stage, the aforementioned questions should be considered.

9.4.2 Project Life Cycle

The typical project life cycle is given in Fig. 9.6. The whole process includes the production stage, R&D and the commissioning of product. In this case, the customer is likely to be in a hospital or a specific patient. The market target is the medical device industry. The planning stage encompasses many aspects: work breakdown structure, personnel, schedule, cost, risks, etc. Then, the product can be developed and product design can be commissioned, followed by manufacturing/production. Note that meeting health/clinical regulatory requirements is also important. Training aims to educate the customer on how to use the machine and after-sales means to maintain the machine in the case of future problems. Some electrical-based components of biomedical devices need improvising such as software upgrading. Evaluation here indicates the assessment of the whole project to conclude and derive valuable lessons, instead of testing the device which should be completed in the execution stage.

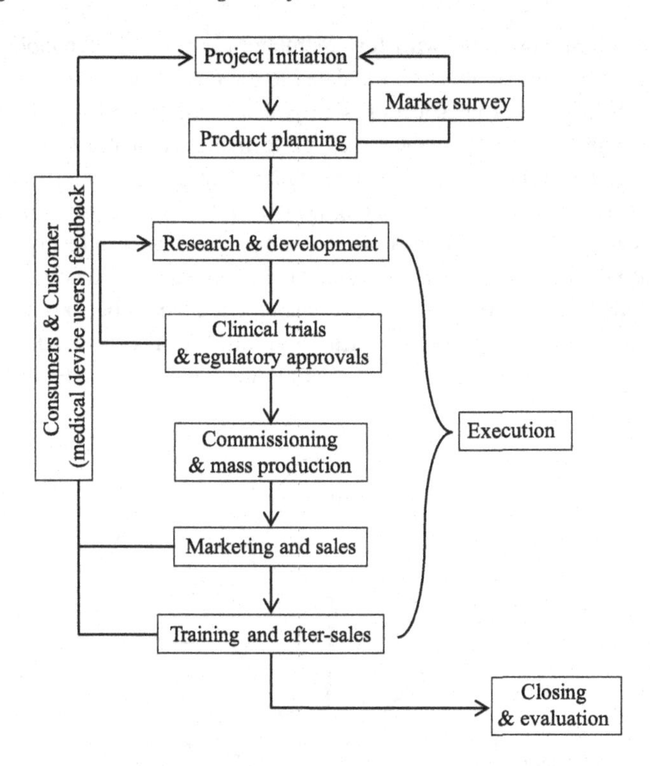

FIGURE 9.6
Project life cycle. A logical sequence of activities is implemented to accomplish the production of a successful biomedical device by meeting specific goals or objectives for its design, production, marketing, etc. Regardless of the biomedical product's scope or complexity, its implementation goes through a series of stages in a project.

9.4.3 Project Management Team

As can be observed in the previous section, a group of specialists or dedicated personnel can be recruited from different departments to form a CFT or committee that serves a specific purpose. Figure 9.7 depicts some of the functions in an organisation that project teams can be formed to look into or to formulate useful projects around it. Likewise, a project management team is usually constructed from members of various departments. An example of a project management organisation structure will be described here in the scope of the engineering at the R&D department. In this project management framework, all design and in-house R&D project or work packages are developed in a streamlined and organised style to generate the final deliverables, which in this aspect, is the biomedical device. The project team is usually kept simple in a hierachical structure and with few members for reasons of enabling efficient communication. Report statuses are escalated to the project leader/manager who can be the R&D head. The project leader is responsible for the delivery of the project or work package, to meet the project key performance indicator (KPI) targets and to build a sustainable and competent team. Support program managers and product managers, who are generally from the engineering or design departments, assume the position of team leader. They lend support to the project leader in terms of managing a group of auxiliary team members to produce components of the deliverables.

Relationships enter into a better shape when the team members know themselves and their roles clearly. For example, a system validation team headed by a team leader typically provides technical quality support in the form of specification/design

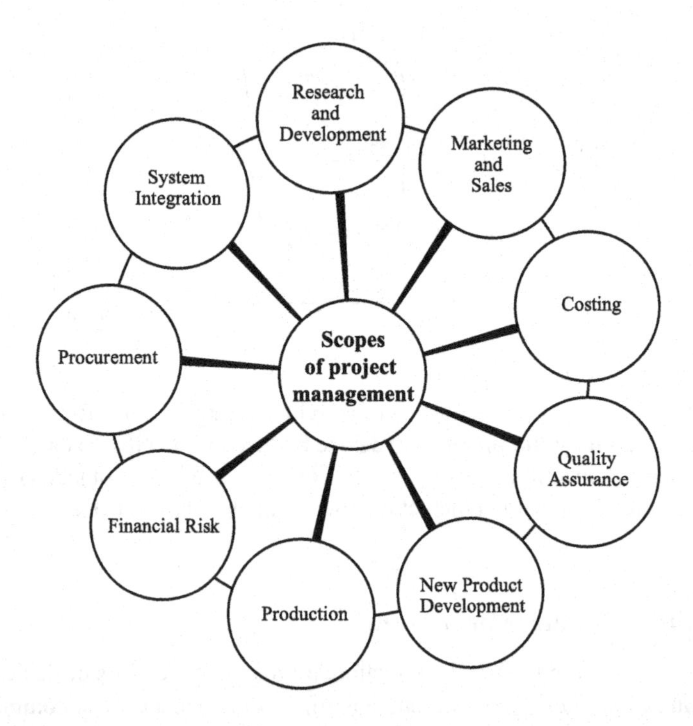

FIGURE 9.7
Project scopes managed by a cross-functional team (CFT) on a product development platform. Project management teams can be formed to look after projects pertaining to specific functions of an organisation's operations that relate to technological product development. Although each of the departments is structured within an organisation to take care of its designated core function, certain projects relating to their specific functions require CFTs. Such teams can be formed from various other departments due to the broad spectrum of expertise required to handle them.

review and traceability support. Test equipment and network build up, as well as the maintenance of production facilities are their responsibility. In the case of automated manufacturing, automation platform and test script development are maintained by this team. It also helps if the project manager is experienced and is capable of effectively evaluating his team members. The importance of good personal rapport among the team members will be emphasised when it develops into a well of goodwill to draw from in case of tension. Let us examine the project management structure with the more elaborate terminologies in Fig. 9.8. The job scopes and responsibilities of the key personnels and members in a project management team are detailed in the following paragraphs.

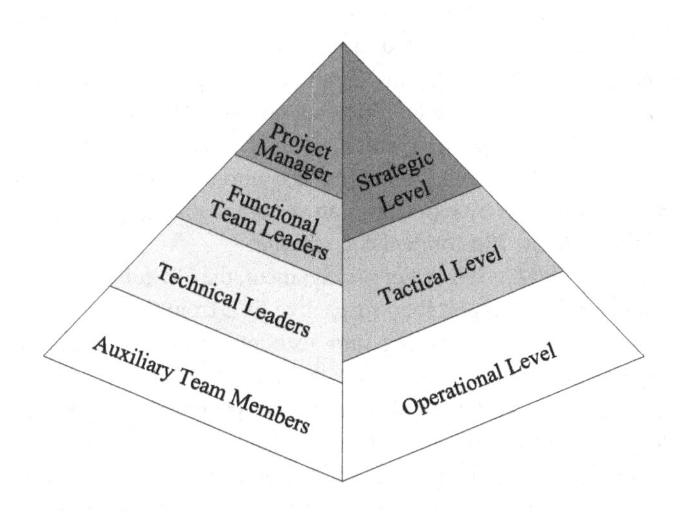

FIGURE 9.8
Organisation chart of a project management team. A project management team can be implemented in every department of the organisation. It comprises the project manager (PM), functional team leaders (FTLs), technical leaders (TLs) and auxiliary team members (ATMs). These personnels form an effective and efficient group of members that work together in a streamlined manner at the Strategic, Tactical and Operational levels to achieve the project deliverables and meet KPI targets.

Project Manager (PM): During project and resource management, coordination by the PM, who is usually the departmental head includes resource assignment priority management, capacity planning, the organisation of monthly/weekly program and weekly project review, as well as the administration of project management tools, e.g., administration of portfolio resources project management (PRPM). In the context of project management, PRPM is a term used to define the framework for analysis and management of projects. The PRPM objective is the optimisation of project implementations to achieve successful product research and development while being constrained by temporal and financial factors. Project leaders need to consider each of the project's costs, the inputs required, expected date of completion, expected outcomes or benefits and the inter-dependencies with the various other projects in the portfolio. The PM is also responsible for ensuring goal alignment, and forming a group of team leaders and technical leaders that pertains to a relatively simple organisational framework. Emphasising an excessive formal structure can reduce the effectiveness of communication. The selection of a PM stresses on someone having stronger interpersonal skills than technical skills.

Functional team leader (FTL): Under the leadership of the PM, the FTL is responsible for leading the competence team in a functional group from a technical perspective. The function can be in design, engineering, testing and evaluation, or marketing and sales. He/she is also in charge of developing the competencies of the team, provide necessary training and help support the deliveries to the projects in terms of quality, time and cost. In addition, the FTL must support the team members (or ATMs) need help. From other technical and non-technical perspectives, he/she can support the other FTLs by reporting statuses or escalating issues. Any important updates are relayed to the PM for decision making.

Technical leader (TL): In project management, the TL guides the internal technical team in the project, support their respective FTL from the technical perspective and provide technical support, taking into account the quality, timescale and cost involved. He/she also reports and escalates the key technical issue/change whenever necessary, and monitor the synchronisation of the specifications, design, implementation and testing. Some examples are given as follows: from the *quality perspective*, lead implementation and product design and visualisation, support re-design and lead technical discussion; from the *time perspective*, support schedule estimation; and from the *cost perspective*, support effort estimation.

Auxiliary team members (ATMs): The lowest echelon of the hierachical structure is made up of ATMs, which is formed by junior executives from various departments such as engineering, marketing and sales, procurement and manufacturing within the organisation. ATMs are selected to join the project team based on their technical expertise and will be guided by the TL to complete sub-tasks of the entire project.

9.5 Summary

In the health care industry, new problems consistently emerge which require new research and changes in medical technologies to meet the needs of their consumers. Typically, in relation to biomedical devices, the approach is to solve these problems arising from current clinical/surgical needs to improve the present product design, to devise new products and finally, to discover new utilities for the present or new products. The fundamental objective boils down to financial sustainability. In an organisation manufacturing biomedical devices, the other objectives may be to sell personnel relations, manufacturing processes or product design in addition to the product itself. The decision of the organisation to focus on pure research in the backend of its engineering depends on company costs. The factors influencing pure research by a plant are mainly due to economic implications. For example, investment in new machineries is often weighed against the possibility of developing new processes from a dedicated research to modify the equipment or extend the usability of the equipment.

While the production department strives towards design standardisation, the research engineer or biomedical device expert devises diversified designs with reference to the standard product objectives. The goal is to meet the improved health care and patient-specific needs of consumers and enhance product competence. This also corresponds to increased cost of manufacturing. Additionally, there is also a challenge of competing against existing products that pertains to variable designs under the constraint of process or design patents by the competitor manufacturer. Minimal modification to the product design to reduce costs is a necessity in order to meet the steep price competition. To achieve this, the continued research of manufacturing processes and designs is the crux of new product development. In some cases, the organisation can choose to develop in-house research capabilities or to leverage on the research facility of an ally organisation, which may be a university. A systematic approach of diffusing new knowledge arising from this research throughout the organisation is helpful for keeping all departments updated on new processes related to product engineering and manufacturing. This can be enabled by modern scientific management of the organisation. The end benefit of a good management strategy is the rapid research and development of new products and speedy modification of manufacturing processes in order to meet market competition.

9.6 Questions

1) How are advances in a product design heavily pegged to technology developments?
2) What is the difference between a patent and a trade secret?
3) Name the six constraints of project management.
4) What are the entities of a new product development advisory board?
5) What are the entities of a project management team?
6) List the typical key departments in an engineering organisation.
7) What are the four phases of a typical project management?

Bibliography

1. Kelly JT, Asgharian B, Kimbell JS, Wong BA. Particle deposition in human nasal airway replicas manufactured by different methods. Part 1: Inertial regime particles. *J Aerosol Sci* 38:1063–1071, 2004.

2. Kubota K, Gomi N, Wakita T, Shibuya H, Kakimoto M, Osanai T. Magnetic resonance imaging of the metal clip in a breast: Safety and its availability as a negative marker. *Breast Cancer* 11(1):55–59, 2004.

3. Sun Z, Allen YB, Nadkarni S, Knight R, Hartley DE, Lawrence-Brown MM. CT virtual intravascular endoscopy in the visualization of fenestrated stent-grafts. *J Endovasc Ther* 15(1):42–51, 2008.

4. Sun Z, Chaichana T, Sangworasil M, Tangjitkusolmun S, Allen Y, Hartley D, Lawrence-Brown M. Computed simulation and analysis of hemodynamic changes in abdominal aortic aneurysms treated with fenestrated endovascular grafts. *3rd ISBME (International Symposium on Biomedical Engineering), Bangkok, Thailand*, pp. 441–446, 2008.

5. Wong KKL, Kelso RM, Worthley SG, Sanders P, Mazumdar J, Abbott D. Theory and validation of magnetic resonance fluid motion estimation using intensity flow data. *PLoS ONE* 4(3):e4747, 2009.

6. Shi Y, Zhao Y, Yeo TJ, Hwang NH. Numerical simulation of opening process in a bileaflet mechanical heart valve under pulsatile flow condition. *J Heart Valve Dis* 12(2):245–255, 2003.

7. Inthavong K, Tian ZF, Li HF, Tu JY, Yang W, Xue CL, Li CG. A numerical study of spray particle deposition in a human nasal cavity. *Aerosol Sci Technol* 40(11):1034–1045, 2006.

8. Maier SE, Meier D, Boesiger P, Moser UT, Vieli A. Human abdominal aorta: Comparative measurements of blood flow with MR imaging and multigated Doppler US. *Radiology* 171:487–492, 1989.

9. Powell AJ, Maier SE, Chung T, Geva T. Phase-velocity cine magnetic resonance imaging measurement of pulsatile blood flow in children and young adults: *In vitro* and *in vivo* validation. *Pediatr Cardiol* 21:104–110, 2000.

10. Lotz J, Meier C, Leppert A, Galanski, M. Cardiovascular flow measurement with phase-contrast MR imaging: Basic facts and implementation. *Radiographics* 22:651–671, 2002.

11. Hatle L, Angelsen B. *Doppler Ultrasound in Cardiology: Physical Principles and Clinical Applications*, 2nd edn. Lea and Febiger, Philadelphia, 1982.

12. Dissanayake DW, Al-Sarawi S, Abbott D. Advanced modeling and simulation of wirelessly interrogated valve-less microfluidic devices. In *Proc. of 2009 IEEE Regional Symposium on Micro and Nano Electronics*, pp. 332–336, August 2009.

13. Nisar A, Afzulpurkar N, Mahaisavariya B, Tuantranont A. Multifield analysis of a piezoelectrically actuated valveless micropump. *Sens Transducers J* 94(7):176–195, 2008.

14. König CS, Clark C, Mokhtarzadeh-Dehghan MR. Comparison of flow in numerical and physical models of a ventricular assist device using low- and high-viscosity fluids. *Proc of the Institution of Mechanical Engineers, Part H: J Eng Med* 213(5):423–432, 1999.

15. Anderson JB, Wood HG, Allaire PE, Bearnson G, Khanwilkar P. Computational flow study of the continuous flow ventricular assist device, prototype number 3 blood pump. *Artif Organs* 24(5):377–385, 2000.

16. Hsu C-H. Flow study on a newly developed impeller for a left ventricular assist device. *J Artif Organs* 6(2):92–100, 2003.

17. Behr M, Arora D, Nosé Y, Motomura T. Performance analysis of ventricular assist devices using finite element flow simulation. *Int J Numer Methods Fluids* 46(12):1201–1210, 2004.

18. Ng EY, Zhou W. Application of CFD techniques to a scaled rotary blood pump. *Int J Comput Fluid Dyn* 8:561–569, 2000.

19. Arvand A, Hahn N, Hormes M, Akdis M, Martin M, Reul H. Comparison of hydraulic and hemolytic properties of different impeller designs of an implantable rotary blood pump by computational fluid dynamics. *Artif Organs* 28(10):892–898, 2004.

20. Xia GH, Zhao Y, Yeo JH. Numerical simulation of 3D fluid-structure interaction using an immersed membrane method. *Mod Phys Lett B* 19(28-29):1447–1450, 2005.

21. Chandran KB, Yoganathan AP, Rittgers SE. *Biofluid Mechanics: The Human Circulation*. CRC Press, Taylor & Francis Group, 2006.

22. Govindarajan V, Udaykumar HS, Chandran KB. Two-dimensional simulation of flow and platelet dynamics in the hinge region of a mechanical heart valve. *J Biomech Eng* 131(3):031002, 2009.

23. de Rochefort L, Vial L, Fodil R, Maître X, Louis B, Isabey D, Caillibotte G, Thiriet M, Bittoun J, Durand E, Sbirlea-Apiou G. *In vitro* validation of computational fluid dynamic simulation in human proximal airways with hyperpolarized 3He magnetic resonance phase-contrast velocimetry. *J Appl Physiol* 102(5):2012–2023, 2007.

24. Kim YH, Xu X, Lee JS. The effect of stent porosity and strut shape on saccular aneurysm and its numerical analysis with lattice boltzmann method. *Ann Biomed Eng* 38:22742292, 2010.

25. Takahashi S, Yagi T, Umezu M. Experimental insight into modeling of cerebral aneurysm hemodynamics: Comparison between stereo PIV and CFD. *11th Asian Symposium on Visualization*, Niigata, Japan, 2011.

26. Parodi JC, Palmaz JC, Barone HD. Transfemoral intraluminal graft implantation for abdominal aortic aneurysms. *Ann Vasc Surg* 5:491–499, 1991.

27. Buth J, van Marrewijk CJ, Harris PL, Hop WCJ, Riambau V, Laheij RJF. Outcome of endovascular abdominal aortic aneurysm repair in patients with conditions considered unfit for an open procedure: A report on the EUROSTAR experience. *J Vasc Surg* 35:211–221, 2002.

28. Lobato AC, Quick RC, Vaughn PL, Rodriguez-Lopez J, Douglas M, Diethrich EB. Transrenal fixation of aortic endografts: Intermediate follow-up of a single-center experience. *J Endovasc Ther* 7:273–278, 2000.

29. Bove PG, Long GW, Shanley CJ, Brown OW, Rimar SD, Hans SS, Kitzmiller JW, Bendick PJ, Zelenock GB. Transrenal fixation of endovascular stent-grafts for infrarenal aortic aneurysm repair: Mid-term results. *J Vasc Surg* 37:938–942, 2003.

30. Browne TF, Hartley D, Purchas S, *et al.* A fenestrated covered suprarenal aortic stent. *Eur J Vasc Endovasc Surg* 18:445–449, 1999.

31. Stanley BM, Semmens JB, Lawrence-Brown MM, Goodman MA, Hartley DE. Fenestration in endovascular grafts for aortic aneurysm repair: New horizons for preserving blood flow in branch vessels. *J Endovasc Ther* 8:16–24, 2001.

32. Anderson JL, Berce M, Hartley D. Endoluminal aortic grafting with renal and superior mesenteric artery incorporation by graft fenestration. *J Endovasc Ther* 8:3–15, 2001.

33. Richter GM, Palmaz JC, Noeldge G, Tio F. Relationship between blood flow, thrombosis and neointima in stents. *J Vasc Interventional Radiol* 10:598–604, 1999.

34. Beythien C, Gutensohn K, Bau J, Hamm CW, Kühnl P, Meinertz T, Terres W. Influence of stent length and heparin coating on platelet activation: A flow cytometric analysis in a pulsed floating model. *Thromb Res* 94(2):79–86, 1999.

35. Ranade VV, Hollinger MA. Implants in drug delivery, In: Ranade VV, Cannon JB, eds. *Drug Delivery Systems*, 2nd edn. CRC Press, Florida, 2004, pp. 115–149.

36. Nisar A, Afzulpurkar N, Tuantranont A, Mahaisavariya B. Three dimensional transient multifield analysis of a piezoelectric micropump for drug delivery system for treatment of hemodynamic dysfunctions. *Cardiovasc Eng* 8:203–218, 2008.

37. Lifespan. Lifespan, 2009, Retrieved January 11, 2009, from http://www.lifespan.org, 2009.

38. Tsai NC, Sue CY. Review of MEMS–based drug delivery and dosing systems. *Sens Actuators A* 134:555–564, 2007.

39. Razzacki SZ, Thwar PK, Yang M, Ugaz VM, Burns MA. Integrated microsystems for controlled drug delivery. *Adv Drug Delivery Rev* 56(2):185–198, 2004.

40. Kimbell J, Shroeter JD, Asgharian B, Wong BA, Segal RA, Dickens CJ, Southall JP, Miller FJ. Optimisation of nasal delivery devices using computational models. *Respir Drug Delivery* 9:233–238, 2004.

41. Zwartz GJ, Guilmette RA. Effect of flow rate on particle deposition in a replica of a human nasal airway. *Inhalation Toxicol* 13(2):109–127, 2001.

42. Suman JD, Laube BL, Lin TC, Brouet G, Dalby R. Validity of *in vitro* tests on aqueous spray pumps as surrogates for nasal deposition. *Pharma Res* 19(1):1–6, 2002.

43. Cheng YS, Holmes TD, Gao J, Guilmette RA, Li S, Surakitbanharn Y, Rowlings C. Characterization of nasal spray pumps and deposition pattern in a replica of the human nasal airway. *J Aerosol Med* 14(2):267–280, 2001.

44. Kippax PG, Krarup H, Suman JD. Applications for droplet sizing — manual versus automated actuation of nasal sprays. *Pharma Technol* pages 30–39, 2004.

45. Farina D. Advancing the science of *in vitro* testing and laboratory data management for nasal sprays. *Drug Delivery Technol* 4:1, 2008.

46. Dayal P, Shaik MS, Singh M. Evaluation of different parameters that affect droplet-size distribution from nasal sprays using the Malvern Spraytec. *J Pharm Sci* 93(7):1725–1742, 2004.

47. Guo C and Doub WH. The influence of actuation parameters on *in vitro* testing of nasal spray products. *J Pharm Sci* 95(9):2029–2040, 2006.

48. Guo C, Stine KJ, Kauffman JF, Doub WH. Assessment of the influence factors on *in vitro* testing of nasal sprays using Box-Behnken experimental design. *Eur J Pharm Sci* 35(5):417–426, 2008.

49. Kiris C, Kwak D, Rogers S, Chang I-D. Computational approach for probing the flow through artificial heart devices. *J Biomech Eng* 119:452–460, 1997.

50. Peskin CS. *Flow patterns around heart valves: A digital computer method for solving the equations of motion*. PhD thesis, Albert Einstein College of Medicine, Yeshiva University, 1972.

51. Sun Z, Winder J, Kelly B, Ellis P, Kennedy P, Hirst D. Diagnostic value of CT virtual intravascular endoscopy in aortic stent-grafting. *J Endovasc Ther* 11:13–25, 2004.

52. Sun Z, Chaichana T. Investigation of hemodynamic effect of stent wires on renal arteries in patients with abdominal aortic aneurysms treated with suprarenal stent grafts. *Cardiovasc Interventional Radiol* 32:647–657, 2009.

53. Dissanayake DW, Al-Sarawi S, Lu T, Abbott D. Finite element modelling of SAW device based corrugated micro-diaphragms. *Smart Mater Struct* 18(9): 095030, 2009.

54. Dissanayake DW, Al-Sarawi S, Abbott D. Advanced modeling and simulation of wirelessly interrogated valve-less microfluidic devices. *Proc 2009 IEEE Regional Symp on Micro and Nano Electronics*, pp. 332–336, 2009.

55. Stytz MR, Frieder G, Frieder O. Three-dimensional medical imaging: Algorithms and computer systems. *ACM Comput Surveys* 23:421–499, 1991.

56. Kelley DJ, Farhoud M, Meyer ME, Nelson DL, Ramirez LF, Dempsey RJ, Wolf AJ, Alexander AL, Davidson RJ. Creating physical 3D stereolithograph models of brain and skull. *PLoS ONE* 2:e1119, 2007.

57. Hustinx R, Alavi, A. SPECT and PET imaging of brain tumours. *Neuroimaging Clin North Am* 9:751–766, 1999.

58. Scollan DF, Holmes A, Zhang J, Winslow RL. Reconstruction of cardiac ventricular geometry and fiber orientation using magnetic resonance imaging. *Ann Biomed Eng* 28:934–944, 2000.

59. Schuijf JD, Jukema JW, Bax JJ. CT imaging of the heart. *Netherlands Heart J* 14:440–441, 2006.

60. Saber NR, Gosman AD, Wood NB, Kilner PJ, Charrier CL, Firmin DN. Computational flow modeling of the left ventricle based on *in vivo* MRI data. *Ann Biomed Eng* 29:275–283, 2001.

61. Kublik H and Vidgren MT. Nasal delivery systems and their effect on deposition and absorption. *Adv Drug Delivery Rev* 29:157–177, 1998.

62. Shipper GM, Verhoef J, Merkus WHM. The nasal mucociliary clearance: Relevance to nasal drug delivery. *Pharm Res* 7:807–814, 1991.

63. Lynch DA, Newell JD, Tschomper BA, Cink TM, Newman LS, Bethel R. Uncomplicated asthma in adults: Comparison of CT appearance of the lungs in asthmatic and healthy subjects. *Radiology* 188:829–833, 1993.

64. Filler AG. MR neurography and diffusion tensor imaging: Origins, history & clinical impact. *Nat Precedings*, 2009.

65. Stahlberg F, Sondergaard L, Thomsen C. MR flow quantification with cardiovascular applications: A short review. *Acta Paediatr Suppl* 410:49–56, 1995.

66. Hartiala JJ, Mostbeck GH, Foster E, Fujita N, Dulce MC, Chazouilleres AF, Higgins CB. Velocity-encoded cine MRI in the evaluation of left ventricular diastolic function: Measurement of mitral valve and pulmonary vein flow velocities and flow volume across the mitral valve. *Am Heart J* 125:1054–1066, 1993.

67. Mankovich NJ, Cheeseman AM, Stoker NG. The display of three-dimensional anatomy with stereolithographic models. *J Digital Imaging* 3(3):200–203, 1990.

68. Eppley BL, Kilgo M, Coleman JJ. Cranial reconstruction with computer-generated hard-tissue replacement patient-matched implants: Indications, surgical technique, and long-term follow-up. *Plast Reconstruct Surg* 109:864–871, 2002.

69. Schantz J-T, Hutmacher DW, Brinkmann M, Lam CXF, Wong KM, Lim TC, Chou N, Guldberg RE, Teoh SH. Repair of calvarial defects with customised tissue-engineered bone grafts II: Evaluation of cellular efficiency and efficacy *in vivo*. *Tissue Eng* 9(1):127–139, 2003.

70. Markl M, Schumacher R, Küffer J, Bley TA, Hennig J. Rapid vessel prototyping: Vascular modeling using 3D magnetic resonance angiography and rapid prototyping technology. *Magn Res Mater Phys Biol Med* 18:288–292, 2005.

71. Sodian R, Fu P, Lueders C, Szymanski D, Fritsche C, Gutberlet M, Hoerstrup SP, Hausmann H, Lueth T, Hetzer R. Tissue engineering of vascular conduits: Fabrication of custom-made scaffolds using rapid prototyping techniques. *Thorac Cardiovasc Surg* 53(3):144–149, 2005.

72. Campbell RI, Martorelli M, Lee HS. Surface roughness visualization for rapid prototyping models. *Comput Aided Des* 34:717–725, 2002.

73. Onuh SO, Hon KKB. Optimizing build parameters for improved finish in stereolithography. *Int J Mach Tools Manuf* 38(4):329–392, 1998.

74. Onuh SO, Hon KKB. Improving stereolithography part accuracy for industrial applications. *Int J Adv Manuf Technol* 17(1):61–68, 2001.

75. Kalender W, Sessler W, Klotz E, Vock P. Spiral volumetric CT with single-breath-hold technique, continuous transport and continuous scanner rotation. *Radiology* 176:181–183, 1990.

76. Crawford CR, King KF. Computed tomography scanning with simultaneous patient translation. *Med Phys* 17:967–982, 1990.

77. Iwano S, Imaizumi K, Okada T, Hasegawa Y, Naganawa S. Virtual bronchoscopy-guided transbronchial biopsy for aiding the diagnosis of peripheral lung cancer. *Eur J Radiol* pages 1–5, 2009.

78. Lin OS. Computed tomographic colonography: Hope or hype? *World J Gastroenterol* 28:915–920, 2010.

79. Davis CP, Ladd ME, Romanowski BJ, *et al.* Human aorta: Preliminary results with virtual endoscopy based on three-dimensional MR imaging data sets. *Radiology* 199:37–40, 1996.

80. Neri E, Bonanomi C, Vignali R, *et al.* Spiral CT virtual endoscopy of abdominal arteries: Clinical applications. *Abdom Imaging* 25:59–61, 2000.

81. Rubin GD, Napel S, Leung AN. Volumetric analysis of volumetric data: Achieving a paradigm shift. *Radiology* 200:312–317, 1996.

82. Bartolozzi C, Neri E, Caramella D. CT in vascular pathologies. *Eur Radiol* 8:679–684, 1998.

83. Finkelstein SE, Summers RM, Nguyen DM, *et al.* Virtual bronchoscopy for evaluating malignant tumors of the thorax. *J Thorac Cardiovasc Surg* 123:967–972, 2002.

84. Dachman AH. Diagnostic performance of virtual colonoscopy. *Abdom Imaging* 27:260–267, 2002.

85. Kimura F, Shen Y, Date S, *et al.* Thoracic aortic aneurysm and aortic dissection: New endoscopic mode for three-dimensional CT display of aorta. *Radiology* 198:573–578, 1996.

86. Sun Z, Winder J, Kelly B, *et al.* CT virtual intravascular endoscopy of abdominal aortic aneurysms treated with suprarenal endovascular stent grafting. *Abdom Imaging* 28:580–587, 2003.

87. Unno N, Mitsuoka H, Takei Y, *et al.* Virtual angioscopy using three-dimensional rotational digital subtraction angiography for endovascular assessment. *J Endovasc Ther* 9:529–534, 2002.

88. Glockner JF. Navigating the aorta: MR virtual vascular endoscopy. *Radiographics* 23:e11, 2003.

89. Sun Z. 3D multislice CT angiography in post-aortic stent grafting: A pictorial essay. *Korean J Radiol* 7:205–211, 2006.

90. Sun Z, Winder J, Kelly B, Ellis PK, Kennedy PT, Hirst DG. Diagnostic value of CT virtual intravascular endoscopy in aortic stent grafting. *J Endovasc Ther* 11:13–25, 2004.

91. ODonnell ME, Sun Z, Winder RJ, Ellis PK, Lau LL, Blair PHB. Suprarenal fixation of endovascular aortic stent grafts: Assessment of medium to long term renal function by analysis of juxta-renal stent morphology. *J Vasc Surg* 45:694–700, 2007.

92. Sun Z, ODonnell M, Winder R, Ellis P, Blair P. Effect of suprarenal fixation of aortic stent grafts on renal ostium: Assessment of morphological changes by virtual intravascular endoscopy. *J Endovasc Ther* 14:650–660, 2007.

93. Sun Z. Multislice CT angiography in abdominal aortic aneurysm treated with endovascular stent grafts: Evaluation of 2D and 3D visualisations. *Biomed Imaging Intervention J* 3, 2007.

94. Beier J, Diebold T, Vehse H, *et al.* Virtual endoscopy in the assessment of implanted aortic stents. *Comput Assisted Radiol Surg* 183–188, 1997.

95. Sun Z, Winder J, Kelly B, Ellis PK, Kennedy PT, Hirst DG. Assessment of VIE image quality using helical CT angiography: *In vitro* phantom study. *Comput Med Imaging Gr* 28:3–12, 2004.

96. Sun Z, Gallagher E. Multislice CT virtual intravascular endoscopy for abdominal aortic aneurysm stent grafts. *J Vasc Interventional Radiol* 15:961–970, 2004.

97. Sun Z, Ng C. Dual-source CT angiography in aortic stent grafting: An *in vitro* aorta phantom study of image noise and radiation dose. *Acad Radiol*, 2010.

98. Sun Z. Three-dimensional visualization of suprarenal aortic tent-grafts: Evaluation of migration in midterm follow-up. *J Endovasc Ther* 13:85–93, 2006.

99. Endovas stent-grafts for the treatment of abdominal aortic aneurysms. *NICE technology appraisal guidance 167. www.nice.org.uk.*

100. Desai M, Eaton-Evans J, Hillery C, Bakhshi R, You Z, Lu J, Hamilton G, Seifalian AM. AAA stent-grafts: Past problems and future prospects. *Ann Biomed Eng* 38(4):1259–1275, 2010.

101. Norrgård O, Angquist K-A, Johnson O. Familial aortic aneurysms: Serum concentrations of triglyceride, cholesterol, HDL-cholesterol and (VLDL + LDL)-cholesterol. *Br J Surg* 72(2):113–116, 1985.

102. Ernst CB. Abdominal aortic aneurysm. *N Engl J Med* 328(16):1167, 1993.

103. Lawrence PF, Gazak C, Bhirangi L, Jones B, Bhirangi K, Oderich G, Treiman G. The epidemiology of surgically repaired aneurysms in the United States. *J Vasc Surg* 30(4):632–640, 1999.

104. Hallett Jr JW, Naessens JM, Ballard DJ. Early and late outcome of surgical repair for small abdominal aortic aneurysms: A population-based analysis. *J Vasc Surg* 18(4):684, 1993.

105. Galland RB, Simmons MJ, Torrie EPH. Prevalence of abdominal aortic aneurysm in patients with occlusive peripheral vascular disease. *Br J Surg* 78(10):1259–1260, 1991.

106. May J, White GH, Yu W, Ly CN, Waugh R, Stephen MS, Arulchelvam M, Harris JP. Concurrent comparison of endoluminal versus open repair in the treatment of abdominal aortic aneurysms: Analysis of 303 patients by life table method. *J Vasc Surg* 27(2):213–221, 1998.

107. Cronenwett JL, Krupski WC, Rutherford RB. Abdominal aortic and iliac aneurysms. *En Vascular Surgery*, WB Saunders Company, Philadelphia, 2000.

108. Cherry KJ. Techniques in the management of recurrent aortic aneurysms. *Aneurysms: New findings and treatments*, pages 249–258, 1994.

109. Perkins JM, Magee TR, Hands LJ, Collin J, Galland RB, Morris PJ. Prospective evaluation of quality of life after conventional abdominal aortic aneurysm surgery. *Eur J Vasc Endovasc Surg* 16(3):203, 1998.

110. Parodi JC, Palmaz JC, Barone HD. Transfemoral intraluminal graft implantation for abdominal aortic aneurysms. *Ann Vasc Surg* 5(6):491–499, 1991.

111. Woodburn KR, May J, White GH. Endoluminal abdominal aortic aneurysm surgery. *Br J Surg* 85(4):435–443, 1998.

112. May J, White GH, Yu W, Waugh RC, McGahan T, Stephen MS, Harris JP. Endoluminal grafting of abdominal aortic aneurysms: Causes of failure and their prevention. *J Endovasc Surg* 1(1):44–52, 1994.

113. Lederle FA. Abdominal aortic aneurysm–open versus endovascular repair. *N Engl J Med* 351(16):1677, 2004.

114. Prinssen M, Verhoeven EL, Buth J, *et al.* A randomized trial comparing conventional and endovascular repair of abdominal aortic aneurysms. *N Engl J Med* 14:1607–1618, 2004.

115. Lobato AC, Quick RC, Vaughn PL, Rodrigues-Lopez J, Douglas M, Diethrich EB. Transrenal fixation of aortic endografts: Intermediate follow-up of a single centre experience. *J Endovasc Ther* 7:273–278, 2000.

116. Lau LL, Hakaim AG, Oldenburg WA, Neuhauser B, McKinney JM, Paz-Fumagalli R, Stockland, A. Effect of suprarenal versus infrarenal aortic endograft fixation on renal function and renal artery patency: A comparative study with intermediate follow-up* 1. *J Vasc Surg* 37(6):1162–1168, 2003.

117. Grego F, Frigatti P, Antonello M, Lepidi S, Ragazzi R, Iurilli V, Zucchetta P, Deriu GP. Suprarenal fixation of endograft in abdominal aortic aneurysm treatment: Focus on renal function. *Ann Surg* 240(1):169, 2004.

118. Sun Z. Helical CT angiography of abdominal aortic aneurysms treated with suprarenal stent grafting: A pictorial essay. *Cardiovasc Interventional Radiol* 26(3):290–295, 2003.

119. Sun Z. Three-dimensional visualization of suprarenal aortic stent-grafts: Evaluation of migration in midterm follow-up. *J Inf* 13(1), 2006.

120. Neri E, Bonanomi G, Vignali C, Cioni R, Ferrari M, Petruzzi P, Bartolozzi C. Spiral CT virtual endoscopy of abdominal arteries: Clinical applications. *Abdom Imaging* 25(1):59–61, 2000.

121. Bartolozzi C, Neri E, Caramella D. CT in vascular pathologies. *Eur Radiol* 8(5):679–684, 1998.

122. Finkelstein SE, Summers RM, Nguyen DM, Stewart IV JH, Tretler JA, Schrump DS. Virtual bronchoscopy for evaluation of malignant tumors of the thorax. *J Thorac Cardiovasc Surg* 123(5):967, 2002.

123. Dachman, AH. Diagnostic performance of virtual colonoscopy. *Abdom Imaging* 27(3):260–267, 2002.

124. Sun Z, Winder RJ, Kelly BE, Ellis PK, Kennedy PT, Hirst DG. Diagnostic value of CT virtual intravascular endoscopy in aortic stent-grafting. *J Endovasc Ther* 11(1):13–25, 2004.

125. Sun Z, Winder RJ, Kelly BE, Ellis PK, Hirst DG. CT virtual intravascular endoscopy of abdominal aortic aneurysms treated with suprarenal endovascular stent grafting. *Abdom Imaging* 28(4):580–587, 2003.

126. Sun Z, O'Donnell ME, Winder RJ, Ellis PK, Blair PH. Effect of suprarenal fixation of aortic stent-grafts on the renal artery ostia: Assessment of morphological changes by virtual intravascular endoscopy. *J Endovasc Ther* 14(5):650–660, 2007.

127. Sun Z, Allen YB, Nadkarni S, Knight R, Hartley DE, Lawrence-Brown MMD. CT virtual intravascular endoscopy in the visualization of fenestrated stent-grafts. *J Endovasc Ther* 15(1):42–51, 2008.

128. Browne TF, Hartley D, Purchas S, Rosenberg M, Van Schie G, Lawrence-Brown M. A fenestrated covered suprarenal aortic stent. *Eur J Vasc Endovasc Surg* 18(5):445–449, 1999.

129. Muhs BE, Verhoeven ELG, Zeebregts CJ, Tielliu IFJ, Prins TR, Verhagen HJM, van den Dungen JJAM. Mid-term results of endovascular aneurysm repair with branched and fenestrated endografts. *J Vasc Surg* 44(1):9–15, 2006.

130. Verhoeven ELG, Prins TR, Tielliu IFJ, van den Dungen J, Zeebregts C, Hulsebos RG, Van Andringa de Kempenaer MG, Oudkerk M, Van Schilfgaarde R. Treatment of short-necked infrarenal aortic aneurysms with fenestrated stent-grafts: Short-term results. *Eur J Vasc Endovasc Surg* 27(5):477–483, 2004.

131. Richter GM, Palmaz JC, Noeldge G, Tio F. Relationship between blood flow, thrombus, and neointima in stents. *J Vasc Interventional Radiol* 10(5):598–604, 1999.

132. Beythien C, Gutensohn K, Bau J, Hamm CW, Kühnl P, Meinertz T, Terres W. Influence of stent length and heparin coating on platelet activation: A flow cytometric analysis in a pulsed floating model. *Thromb Res* 94(2):79–86, 1999.

133. Peacock J, Hankins S, Jones T, Lutz R. Flow instabilities induced by coronary artery stents: Assessment with an *in vitro* pulse duplicator. *J Biomech* 28(1):17–26, 1995.

134. Wilson N, Wang K, Dutton R, Taylor C. A software framework for creating patient specific geometric models from medical imaging data for simulation based medical planning of vascular surgery. In *Medical Image Computing and Computer-Assisted Intervention–MICCAI 2001*, pp. 449–456. Springer, 2010.

135. Bekkers EJ, Taylor CA. Multiscale vascular surface model generation from medical imaging data using hierarchical features. *IEEE Trans Med Imaging* 27(3):331–341, 2008.

136. Zarins CK, Taylor CA. Endovascular device design in the future: Transformation from trial and error to computational design. *J Endovasc Ther* 16(Suppl I):12–21, 2009.

137. Howell BA, Kim T, Cheer A, Dwyer H, Saloner D, Chuter TAM. Computational fluid dynamics within bifurcated abdominal aortic stent-grafts. *J Endovasc Ther* 14(2):138–143, 2007.

138. Liffman K, Lawrence-Brown MMD, Semmens JB, Bui A, Rudman M, Hartley DE. Analytical modeling and numerical simulation of forces in an endoluminal graft. *J Endovasc Ther* 8(4):358–371, 2001.

139. Vignon-Clementel IE, Alberto Figueroa C, Jansen KE, Taylor CA. Outflow boundary conditions for three-dimensional finite element modeling of blood flow and pressure in arteries. *Comput Methods Appl Mech Eng* 195(29-32):3776–3796, 2006.

140. Figueroa CA, Vignon-Clementel IE, Jansen KE, Hughes TJR, Taylor CA. A coupled momentum method for modeling blood flow in three-dimensional deformable arteries. *Comput Methods Appl Mech Eng* 195(41-43):5685–5706, 2006.

141. Sun Z, Chaichana T. Investigation of the hemodynamic effect of stent wires on renal arteries in patients with abdominal aortic aneurysms treated with suprarenal stent-grafts. *Cardiovasc Interventional Radiol* 32(4):647–657, 2009.

142. Rogers C, Edelman ER. Endovascular stent design dictates experimental restenosis and thrombosis. *Circulation* 91(12):2995, 1995.

143. Kuntz RE, Baim DS. Defining coronary restenosis. Newer clinical and angiographic paradigms. *Circulation* 88(3):1310, 1993.

144. Frauenfelder T, Lotfey M, Boehm T, Wildermuth S. Computational fluid dynamics: Hemodynamic changes in abdominal aortic aneurysm after stent-graft implantation. *Cardiovasc Interventional Radiol* 29(4):613–623, 2006.

145. Li Z, Kleinstreuer C. Blood flow and structure interactions in a stented abdominal aortic aneurysm model. *Med Eng Phys* 27(5):369–382, 2005.

146. Scott RAP, Chutter TAM. Clinical endovascular placement of bifurcated graft in abdominal aortic aneurysms without laparotomy. *Lancet*, page 343: 413, 1994.

147. Chutter TA, Green RM, Ouriel K, *et al.* Transfemoral endovascular aortic graft placement. *J Vasc Surg* 18:185–197, 1993.

148. Croenwett JL, Murphy TF, Selenock GB, *et al.* Actuarial analysis of variables associated with rupture of small abdominal aortic aneurysms. *Surgery* 98:472–483, 1985.

149. Wisselink W, Hollier LH. Principles in the technique of endovascular grafting. In: Yao JST, Pearce WT, eds. *Aneurysms: New Findings and Treatments.* Norwalk, 1994.

150. Kaufman JA, Geller SC, Brewer DC, *et al.* Endovascular repair of abdominal aortic aneurysms: Current status and future directions. *Am J Roentgenol* 175:289–302, 2000.

151. Marin ML, Parsons RE, Hollier LH, *et al.* Impact of transrenal aortic endograft placement on endovascular graft repair of abdominal aortic aneurysms. *J Vasc Surg* 28:638–646, 1998.

152. Birch PC, Start RD, Whitebread T, *et al.* The effect of crossing porcine renal artery ostia with various endovascular stents. *Eur J Vasc Endovasc Surg* 17:185–190, 1999.

153. Malina M, Lindh M, Ivancev K, *et al.* The effects of endovascular aortic stents placed across the renal arteries. *Eur J Vasc Endovasc Surg* 13:207–213, 1997.

154. Ferko A, Krajina A, Jon B, *et al.* Juxtarenal aortic aneurysm: Endoluminal transfemoral repair? *Eur Radiol* 7:703–707, 1997.

155. Greenberg RK, Haulon S, Lyden SP, *et al.* Endovascular management of juxtarenal aneurysms with fenestrated endovascular grafting. *J Vasc Surg* 39:279–287, 2004.

156. Muhs BE, Verhoeven EL, Zeebregts CJ, *et al.* Mid-term results of endovascular aneurysm repair with branched and fenestrated endografts. *J Vasc Surg* 44:9–15, 2006.

157. Armon MP, Yusuf SW, Whitaker SC, *et al.* Influence of abdominal aortic aneurysm size on the feasibility of endovascular repair. *J Endovasc Surg* 4:279–283, 1997.

158. Balm R, Stokking R, Kaatee R, *et al.* Computed tomographic angiographic imaging of abdominal aortic aneurysms: Implications for transfemoral endovascular aneurysm management. *J Vasc Surg* 26:232–237, 1997.

159. Sun Z. 3D visualisation of suprarenal aortic tent-grafts: Evaluation of migration in mid-term follow-up. *J Endovasc Ther* 13:85–93, 2006.

160. Illig KA, Green RM, Ouoriel K, *et al.* Fate of the proximal aortic cuff: Implications for endovascular aneurysm repair. *J Vasc Surg* 26:492–501, 1997.

161. White GH, Yu W, May J, *et al.* Endoleak as a complication of endoluminal grafting of abdominal aortic aneurysms: Classification, incidence, diagnosis and management. *J Endovasc Surg* 4:152–168, 1997.

162. Parodi JC, Barone A, Piraino R, *et al.* Endovascular treatment of abdominal aortic aneurysms: Lessons learned. *J Endovasc Surg* 4:102–110, 1997.

163. Chuter TA, Green RM, Ouriel K, *et al.* Infrarenal aortic aneurysm structure: Implications for transfemoral repair. *J Vasc Surg* 20:44–50, 1994.

164. Lumsden AB, Allen RC, Chaikof EL, *et al.* Delayed rupture of aortic aneurysms following endovascular stent grafting. *Am J Surg* 170:174–178, 1995.

165. May J, White GH, Yu W, *et al.* Surgical management of complications following endoluminal grafting of abdominal aortic aneurysms. *Eur J Vasc Endovasc Surg* 10:51–59, 1995.

166. Rydberg J, Kopecky KK, Lalka SG, *et al.* Stent grafting of abdominal aortic aneurysms: Pre- and postoperative evaluation with multislice helical CT. *J Comput Assisted Tomogr* 25:580–586, 2001.

167. Blankensteijn J, Lindenburg F, Van Der Graaf Y, *et al.* Influence of study design on reported mortality and morbidity rates after abdominal aortic aneurysm repair. *Br J Surg* 85:1624–1630, 1998.

168. Walker S, Macierewicz J, MacSweeney S, *et al.* Mortality rates following endovascular repair of abdominal aortic aneurysms. *J Endovasc Surg* 6:233–238, 1999.

169. Greenhalgh RM, Brown LC, Kwong GP, *et al.* Comparison of endovascular aneurysm repair with open repair in patients with abdominal aortic aneurysm (evar trial1), 30-day operative mortality results: Randomised controlled trial. *Lancet* 364:843–848, 2004.

170. ODonnell ME, Sun Z, Winder J, Lau LL, Ellis PK, Blair PH. Suprarenal fixation of endovascular aortic stent grafts: Assessment of medium-term to long-term renal function by analysis of juxtarenal stent morphology. *J Vasc Surg* 45:694–700, 2007.

171. England A, Butterfield JS, Ashleigh RJ. Incidence and effect of bare suprarenal stent struts crossing renal ostia following evar. *Eur J Vasc Endovasc Surg* 32:523–528, 2006.

172. Cayne NS, Rhee SJ, Veith FJ, *et al.* Does transrenal fixation of aortic endografts impair renal function? *J Vasc Surg* 38:639–644, 2003.

173. Burks Jr JA, Faries PL, Gravereaux EC, Hollier LH, Marin ML. Endovascular repair of abdominal aortic aneurysms: Stent-graft fixation across the visceral arteries. *J Vasc Surg* 35:109–13, 2002.

174. Kichikawa K, Uchida H, Maeda M, *et al.* Aortic stent-grafting with transrenal fixation: Use of newly designed spiral z-stent endograft. *J Endovasc Ther* 7:184–91, 2000.

175. Kramer SC, Seifarth H, Pamler R, *et al.* Renal infarction following endovascular aortic aneurysm repair: Incidence and clinical consequences. *J Endovasc Ther* 9:98–102, 2002.

176. Sun Z, ODonnell M, Winder R, Ellis P, Blair P. Effect of suprarenal fixation of aortic stent grafts on renal ostium: Assessment of morphological changes by virtual intravascular endoscopy. *J Endovasc Ther* 14:650–660, 2007.

177. Greenberg RK, Haulon S, Lyden SP, *et al.* Endovascular management of juxtarenal aneurysms with fenestrated endovascular grafting. *J Vasc Surg* 39:279–287, 2004.

178. Sun Z, Mwipatayi B, Semmens JB, Lawrence-Brown MMD. Short to mid-term outcomes of fenestrated endovascular grafts in the treatment of abdominal aortic aneurysms: A systematic review. *J Endovasc Ther* 13:747–753, 2006.

179. Semmens JB, Lawrence-Brown MD, Hartley DE, Allen YB, Green R, Nadkarni S. Outcomes of fenestrated endografts in the treatment of abdominal aortic aneurysms in Western Australia (1997–2004). *J Endovasc Ther* 13:320–329, 2006.

180. Sun Z, Allen Y, Nadkarni S, Knight R, Hartley D, Lawrence-Brown MMD. CT virtual intravascular endoscopy in the visualization of fenestrated endovascular grafts. *J Endovasc Ther* 15:42–51, 2008.

181. Sun Z, Allen Y, Mwipatayi B, Hartley D, Lawrence-Brown MMD. Multislice CT angiography in the follow-up of fenestrated endovascular grafts: Effect of slice thickness on 2D and 3D visualization of the fenestrated stents. *J Endovasc Ther* 15:417–426, 2008.

182. Peacock J, Hankins S, Jones T, Lutz R. Flow instabilities induced by coronary artery stents: Assessment with an *in vitro* pulse duplicator. *J Biomech* 28:17–26, 1995.

183. Chong CK, How TV. Flow patterns in an endovascular stent-graft for an abdominal aortic aneurysm repair. *J Biomech* 37(1):89–97, 2004.

184. Fillinger MF, Raghavan ML, Marra SP, Cronenwett JL, Kennedy FE. *In vivo* analysis of mechanical wall stress and abdominal aortic aneurysm rupture risk. *J Vasc Surg* 36(3):589–597, 2002.

185. Frauenfelder T, Lotfey M, Boehm MT, Wildermuth TS. Computational fluid dynamics: Hemodynamic changes in abdominal aortic aneurysm after stent-graft implantation. *Cardiovasc Interventional Radiol* 29(4):613–623, 2006.

186. Li Z, Kleinstreuer C. Analysis of biomechanical factors affecting stent-graft migration in an abdominal aortic aneurysm model. *J Biomech* 39:2264–2273, 2006.

187. Gawenda M, Knez P, Winter S, *et al.* Endotension is influenced by wall compliance in a latex aneurysm model. *Eur J Vasc Endovasc Surg* 27:45–50, 2004.

188. Morris L, Delassus O, Walsh M, McGloughlin T. A mathematical model to predict the *in vivo* pulsatile drag forces acting on bifurcated stent grafts used in endovascular treatment of abdominal aortic aneurysms (AAA). *J Biomech* 37:1087–1095, 2004.

189. Scotti CM, Finol EA, Viswanathan S, *et al.* Computational fluid dynamics and solid mechanics analyses of a patient-specific AAA pre- and post-EVAR. In *ASME IMECE: Bioengineering*, 2004.

190. Li Z, Kleinstreuer C. Blood flow and structure interactions in a stented abdominal aortic aneurysm model. *Med Eng Phys* 27:369–383, 2005.

191. Liffman K, Lawrence-Brown MD, Semmens JB, *et al.* Suprarenal fixation: Effect on blood flow of an endoluminal stent wire across an arterial orifice. *J Endovasc Ther* 10:260–274, 2003.

192. Sun Z, Winder J, Kelly B, Ellis P, Hirst D. CT virtual intravascular endoscopy of abdominal aortic aneurysms treated with suprarenal endovascular stent grafting. *Abdom Imaging* 28:580–587, 2003.

193. Borghi A, Wood N, Mohiaddin R, Xu X. Fluid-solid interaction simulation of flow and stress pattern in thoracoabdominal aneurysms: A patient-specific study. *J Fluid Struct* 2:270–280, 2007.

194. Sun Z, Mwipatayi B, Allen Y, Hartley D, Lawrence-Brown M. Multislice CT virtual intravascular endoscopy in the evaluation of fenestrated stent graft repair of abdominal aortic aneurysms: A short-term follow-up. *Aust NZ J Surg* 79:836–840, 2009.

195. Sun Z. Helical CT angiography of fenestrated stent grafting of abdominal aortic aneurysms. *Biomed Imaging Intervention J* 2:1–9, 2009.

196. Sun Z and Chaichana T. Fenestrated stent graft repair of abdominal aortic aneurysm: Hemodynamic analysis of effect of fenestrated stents on renal arteries. *Korean J Radiol* 11:95–106, 2010.

197. Lawrence-Brown M, Sun Z, Semmens JB, Liffman K, Sutalo I, Hartley D. Type II endoleaks: When is intervention indicated and the index of suspicion for type I or III? *J Endovasc Ther* 16:106–118, 2009.

198. Chong CK, How TV, Gilling-Smith GL, Harris PL. Modeling endoleaks and collateral reperfusion following endovascular AAA exclusion. *J Endovasc Ther* 10:424–432, 2003.

199. Milner R, Ruurda JP, Blankensteijn JD. Durability and validity of a remote, miniaturized pressure sensor in an animal model of abdominal aortic aneurysm. *J Endovasc Ther* 11:372–377, 2004.

200. Springer F, Gunther RW, Schmitz-Rode T. Aneurysm sac pressure measurement with minimally invasive implantable pressure sensors: An alternative to current surveillance regimes after evar? *Cardiovasc Interventional Radiol* 31:460–467, 2008.

201. Milner R, Kasirajan K, Chaikof EL. Future of endograft surveillance. *Semin Vasc Surg* 19:75–82, 2006.

202. Perry RH, Green D. *Chemical Engineers Handbook.* McGraw-Hill, New York, 1984.

203. Saloman L. *Two-Phase Flow in Complex Equipment.* John Wiley and Sons, Inc., New York, 1999.

204. Hörschler I, Meinke M, Schröder W. Numerical simulation of the flow field in a model of the nasal cavity. *Comput Fluids* 32(1):39–45, 2003.

205. Zamankhan P, Ahmadi G, Wang Z, Hopke PH, Cheng YS, Su WC, Leonard D. Airflow and deposition of nanoparticles in a human nasal cavity. *Aerosol Sci Technol* 40:463–476, 2006.

206. Zhang Y, Finlay WH, Matida EA. Particle deposition measurements and numerical simulation in a highly idealized mouth-throat. *J Aerosol Sci* 35:789–803, 2004.

207. Inthavong K, Tian ZF, Tu JY, Yang W, Xue C. Optimising nasal spray parameters for efficient drug delivery using computational fluid dynamics. *Comput Biol Med* 38(6):713–726, 2008.

208. Inthavong K, Wen J, Tian ZF, Tu JY. Numerical study of fibre deposition in a human nasal cavity. *J Aerosol Sci* 39(3):253–265, 2008.

209. Schroeter JD, Kimbell JS, Asgharian B. Analysis of particle deposition in the turbinate and olfactory regions using a human nasal computational fluid dynamics model. *J Aerosol Med* 19(3):301–313, 2006.

210. Shi H, Kleinstreuer C, Zhang Z. Laminar airflow and nanoparticle or vapor deposition in a human nasal cavity model. *J Biomech Eng* 128(5):697–706, 2006.

211. Shi H, Kleinstreuer C, Zhang Z. Modeling of inertial particle transport and deposition in human nasal cavities with wall roughness. *J Aerosol Sci* 38(4):398–419, 2007.

212. Longest PW, Xi J. Effectiveness of direct lagrangian tracking models for simulating nanoparticle deposition in the upper airways. *Aerosol Sci Technol* 41(4):380–397, 2007.

213. Shi H, Kleinstreuer C, Zhang Z. Dilute suspension flow with nanoparticle deposition in a representative nasal airway model. *Phys Fluids* 20(013301):1–23, 2008.

214. Robert GH. Forced inspiratory nasal flow volume curves: A simple test of nasal airflow. *Mayo Clin Proc* 76:990–994, 2001.

215. Proctor DF. *The Upper Airway, in the Nose.* Elsevier Biomedical Press, New York, 1982.

216. Keyhani K, Scherer PW, Mozell MM. Numerical simulation of airflow in the human nasal cavity. *J Biomech Eng* 117:429–441, 1995.

217. Kimbell JS, Segal RA, Asgharian B, Wong BA, Schroeter JD, Southhall JP, Dickens CJ, Brace J, Miller FJ. Characterization of deposition from nasal spray devices using computational fluid dynamics model of human nasal passages. *J Aerosol Med* 20:59–74, 2007.

218. Nisar A, Afzulpurkar N, Mahaisavariya B, Tuantranont A. MEMS-based micropumps in drug delivery and biomedical applications. *Sens Actuators B* 130(2):917–942, 2008.

219. Li Y, Shawgo RS, Tyler B, Henderson PT, Vogel JS, Rosenberg A, Storm PB, Langer R, Brem H, Cima MJ. *In vivo* release from a drug delivery MEMS device. *J Controlled Release* 100(2):211–219, 2004.

220. Nguyen NT, Huang X, Chuan TK. MEMS-micropumps: A review. *Trans ASME J Fluids Eng* 124:384–392, June 2002.

221. Squires T, Quake S. Microfluidics: Fluid physics at the nanoliter scale. *Rev Mod Phys* 77(3), 2005.

222. Nisar A, Afzulpurkar N, Mahaisavariya B, Tuantranont A. Multifield analysis of a piezoelectrically actuated valveless micropump. *Sens Transducers J* 94(7):176–195, July 2008.

223. Pitz I, Hall LT, Hansen HJ, Varadan VK, Bertram CD, Maddocks S, Enderling S, Saint D, Al-Sarawi SF, Abbott D. Trade–offs for wireless transcutaneous RF communication in biotelemetric applications. *Proc of SPIE–Biomedical Applications of Micro- and Nanoengineering*, 4937:307–318, November 2002.

224. Cui Q, Liu C, Zha XF. Simulation and optimization of a piezoelectric micropump for medical applications. *Int J Adv Manuf Technol* 36(5):516–524, March 2008.

225. Lee S, Kim KJ. Design of IPMC actuator-driven valve-less micropump and its flow rate estimation at low Reynolds numbers. *Smart Mater Struct* 15(4):1103–1109, 2006.

226. Dong-Ho Ha, Van Phuoc Phan, Nam Seo Goo, and Cheol Heui Han. Three-dimensional electro-fluid-structural interaction simulation for pumping performance evaluation of a valveless micropump. *Smart Mater Struct* 18(10):104015, 2009.

227. Dissanayake DW, Tikka AC, Al-Sarawi S, Abbott D. Radio frequency controlled microvalve for biomedical applications. *Proc of SPIE–Smart Materials IV*, 6413:Article 64130D: 1–13, 2007.

228. Tikka AC, Al-Sarawi S, Abbott D, Wong MSK, Schutz JD. Improving the security and actuation of wireless controlled microvalve. *Proc of SPIE–Smart Structures, Devices, and Systems III*, 6414, January 2007.

229. Lee B, Kim ES. Analysis of partly–corrugated rectangular diaphragms using the Rayleigh-Ritz method. *J Microelectromech Syst* 9(3):399–406, September 2000.

230. Bao MH. Basic mechanics of beams and diaphragm structures. In: Bao, MH. *Micro Mechanical Transducers: Pressure Sensors, Accelerometers, and Gyroscopes (Handbook of Sensors and Actuators)*. Elsevier, New York, 2000, pp. 23–87.

231. ANSYS Incorporation. ANSYS homepage. Retrieved November 12, 2009, from http://www.ansys.com/.

232. ANSYS Incorporation. Retrieved November 12, 2009, from ANSYS Help Guide–V.11, http://www.kxcad.net/ansys/ANSYS/ansyshelp/index.htm.

233. White FM. Viscous flow in ducts. In: *Fluid Mechanics*, 4th edn. McGraw-Hill, New York, 1999, pp. 325–424.

234. Yao Q, Xu D, Pan LS, Melisa Teo AL, Ho WM, Peter Lee VS, Shabbir M. CFD simulations of flows in valveless micropumps. *Eng Appl Comput Fluid Mech* 1(3):181–188, 2007.

235. Harvard business review on strategic alliances. Harvard Business School Publishing Corporation, USA, 2002.

236. Gerosa G, Marco F, Casarotto D, Bottio T. Searching for a correct method of evaluation for valve prosthesis performance. *J Heart Valve Dis* 13(Suppl. 1): S1–S3, 2004.

237. Medtronic, Inc. Retrieved November 12, 2009, from http://www.medtronic.com.

238. Eisenberg SR, Nelson RR. Public *vs.* proprietary science: A fruitful tension? *Daedalus* 131:89–101, 2002.

239. Bessen J, Maskin E. Sequential innovation, patents, and imitation. *Massachusetts Institute of Technology Working Paper, Department of Economics, http://www.researchoninnovation.org/patent.pdf*, pp. 2, 2000.

Answers to Chapter Questions

Chapter 1

1) They are (a) measurement, (b) verification, (c) evaluation, (d) modification and (e) optimisation.
2) Pure research seeks for the truth without a predefined utility at the start, whereas applied research pertains to the development of applications to solve problems with the motivation of addressing needs.
3) (a) Autonomous: design realisation independent from its other existing innovations, whereby a new design rule to increase the effectiveness of the device or its level of safety can be developed without redesign of the entire product and (b) systematic: design realisation in conjunction with its related complementary innovations, whereby new design updates requires the need to reformulate the old design rules and redefine the design of the product.
4) They are (a) experimental modelling (particle image velocimetry, phase Doppler particle analyser, laser Doppler anemometry, hot-film anemometry) and (b) analytical modelling (analytical geometry, biofluid mechanics).
5) Experimental flow velocimetry can be based on (a) optical, (b) magnetic resonance and (c) ultrasonic imaging systems.

Chapter 2

1) They are (a) drug delivery mechanism, (b) surgical restoration, (c) medical-assisting device (external support device, surgically implanted device) and (d) diagnostic device.
2) Aneurysm is defined as the focal dilatation in the arterial wall and is commonly caused by atherosclerotic disease of blood vessels. Treatment: insert an aneurysm stent to embolise it or exclude it from systemic blood circulation, so that the aneurysm will gradually shrink (due to reduced pressure) and eventually become smaller.
3) They are (a) drug delivery, drug infusion and dispensing, (b) controlled contraception for males and (c) bladder control.

4) They are (a) electrostatic, (b) piezoelectric, (c) electromagnetic, (d) thermo-pneumatic and (e) shape memory alloy.
5) A large proportion of the drug particles deposited in the anterior regions of the nasal vestibule, and attributed to the sprayed particles existing in a high inertial regime.
6) Excessive deposition of drug particles in the upper airways will cause diminished therapeutic effects in the lung or local side effects in the upper conducting airways.

Chapter 3

1) They are (a) laser additive manufacturing (LAM), (b) selective laser melting (SLM), (c) laser metal deposition (LMD), (d) fused deposition modelling (FDM) and (e) laminated object manufacturing (LOM).
2) They are (a) computed tomography (CT), (b) ultrasound magnetic resonance imaging (MRI), (c) positron emission tomography and (d) single photon emission CT (SPECT).
3) Some specific quantitative functions are the (a) measurement of flow in cardiovascular system using the phase contrast MRI and (b) measurement of local micro-structural characteristics using diffusion MRI.
4) The triangulation is used.
5) Some technical limitations are (a) fidelity of the product depends on the STL file, which in turn dependents on the initial 2D slice spacing and (b) non-smooth object surfaces.

Chapter 4

1) This is a new method of image data visualisation using computer processing of 3D image datasets (such as CT or magnetic resonance scans).
2) They are (a) CT angiography (CTA), (b) magnetic resonance angiography (MRA) and (c) digital subtraction angiography (DSA).
3) The two aspects are (a) providing unique information about the configuration and number of suprarenal stent wires crossing the renal artery ostia and (b) assessing the morphological change of the renal artery ostia following suprarenal stent graft placement.
4) An optimal threshold is important to produce VIE that are free of artefacts.
5) It is due to the improved spatial and temporal resolution of the technique.

Chapter 5

1) CT is limited to visualising anatomical or structural changes, and is lacking in the ability to provide information about the haemodynamic impact of the stent grafts on aortic branches after the implantation of the stent grafts.

2) CFD allows for an early detection of abnormal changes and improves the understanding of the treatment outcomes of endovascular repair, so that the prevention of potential complications and better patient management can be achieved.

3) Spiral CT has improved the performance of CT by converting a 2D modality into true 3D imaging, and hence enabling the development of new applications involving volumetric imaging, such as CT angiography (CTA).

4) They are (a) the blood flow in the circulatory system, (b) the mechanical behaviour of the vessel wall and surrounding tissues under pulsatile and non-pulsatile loading and (c) the mechanical behaviour of the device.

5) VIE can offer a clear 3D intraluminal view of the stent wires and their position relative to the renal ostia.

Chapter 6

1) They are the (a) placement of an uncovered top stent over the rental artery ostia and (b) use of fenestrated stent grafts.

2) Suprarenal fixation is a modification of the commonly used infrarenal fixation, which is evolved to establish a more secure proximal fixation in patients with unfavorable proximal neck anatomy.

3) The principles of fenestration are to preserve blood flow to renal or visceral vessels and enhance stability by inserting stents into side branches to produce a durable relationship between the stent graft fenestration and the branch ostium.

4) They are (a) interference with renal blood flow or renal function, (b) decreased cross-sectional area of the renal ostium and (c) a biological response of the aorta to the aortic stents.

5) The Navier–Stokes equation.

Chapter 7

1) They are (a) initial boundary conditions, (b) numerical technique to be used, (c) mesh quality and (d) geometry reconstruction.

2) They are (a) nasal cavity geometry and (b) flow rate.

3) The function of a nasal spray delivery device is to atomise the liquid formulation into a fine spray that is made up of small micron-sized particles.
4) The monodispersed particle deposition percentage increases with particle size for a given insertion angle.
5) They are (a) redirecting the release point of the spray (the insertion angle) to align with flow streamlines, (b) controlling the particle size distribution and (c) controlling the particle's initial velocity.

Chapter 8

1) Hypertension is due to (a) the heart pumping blood with excessive force or (b) the body's smaller blood vessels (arterioles) becoming more narrow. Both conditions cause the blood flow to exert more pressure against the vessel walls.
2) They are (a) beta-blockers, (b) angiotensin-converting enzyme (ACE) and (c) angiotensin receptor blockers (ARBs).
3) They are (a) electric potential difference between the plates, (b) gab between the plates, (c) dielectric coefficient of the medium between plates and (d) effective plate area.
4) They are (a) input IDT, (b) output IDT and (c) piezoelectric substrate.
5) The SAW device provides a low-powered RFID type, passive wireless interrogation capability for a micropump.

Chapter 9

1) A technological product slowly progresses into the more advanced stages of its development by the incorporation of new technological components, which results in the incremental incorporation of new scientific principles into their design.
2) Patents release the product blueprints to the public, but restrict the exclusive rights to develop the technology for the other parties. In contrast to patents, trade secrets restrict the disclosure of an organisation's technological innovations and inventions to the general public.
3) They are (a) quality, (b) schedule, (c) budget, (d) scope, (e) resources and (f) risks.
4) They are (a) engineering/production team, (b) manufacturing team (c) cost analysis team and (d) quality assurance team.
5) They are (a) project manager (PM), (b) functional team leaders (FTLs) (c) technical leaders (TLs), and (d) auxiliary team members (ATMs).

6) The typical departments are (a) engineering, (b) manufacturing, (c) procurement, (d) sales and (e) finance.
7) The four phases are (a) initiate, (b) plan, (c) execute and (d) closing and evaluation.

Index

abdominal aortic aneurysm (AAA), 55, 69, 79
aneurysm, 9
  aorta, 43, 56, 71
  aortoiliac, 57
  endovascular repair, 57
  juxta-suprarenal, 57
  neck, 57, 72
  neck diameter, 73
  thoracoabdominal, 57
angiography, 38

bending stiffness, 116
Bio-MEMS, 12, 19, 111
  contraception, 114
  electrostatic field, 115
  flow modulation, 115
  lab-on-a-chip, 115
  microdiaphragm, 116
biomedical device, 1, 7
  biocompatibility, 65
  categorisation, 7
  development, 2
  diagnosis, 8
  drug delivery, 8
  mechanical part replacement, 8
  medical-assisting, 8
  microfluidic, 117
  pacemaker, 130
  production, 128
  prosthetic heart valves, 17
    mechanical heart valve, 18
    tissue heart valve, 18
  R&D, 141
  surgical restoration, 8
boundary condition, 97

common iliac arteries, 43, 50
computational fluid dynamics (CFD), 2, 5, 62, 86, 89, 112, 119
computed tomography (CT), 18, 26, 37, 58
  attenuation, 43
  scanner, 51
  spiral CT, 37, 41
computer tomography, 24
computer-aided design (CAD), 20, 80
  CAD software, 18
conceptualisation, 2, 129

deposition pattern, 92, 98, 101
design innovation, 2
  autonomous, 2
  systemic, 2

electrostatic force, 117
endovascular aneurysm repair, 66
endovascular stent graft, 10
experimental flow, 4

fenestrated stent graft, 26, 58, 72, 73, 84
  wire thickness, 86
finite element analysis, 5, 112
flow rate, 96
fluid–structure interaction (FSI), 19, 64, 114, 120
fused deposition modelling, 32

gamma scintigraphy imaging, 90

Hounsfield unit (HU), 43, 46, 49, 79
hypertension, 113

laminar flow, 97
laminated object manufacturing, 32
laser additive manufacturing, 30
  build layer thickness, 31
laser metal deposition, 32

magnetic resonance angiography, 41
magnetic resonance imaging (MRI), 28
  velocity-encoded, 5
manufacturer, 128
manufacturing
  new product, 137
  Six Sigma, 139
  systematic, 134
market survey, 129
mass continuity, 117
material resource planning, 135
medical image segmentation, 26
medical imaging, 4
medical research, 1
mesh, 18, 93
  convergence, 86
  prism, 95
  tetrahedral, 82
microdiaphragm, 121
micron-sized particle, 92
minimally invasive surgery, 57
multi-phase flow, 89
multifield analysis, 124

nasal cavity, 26, 92, 95, 107
  nasal vestibule, 99
  nasopharynx, 101
  nasopharynx outlet, 97
  nostril inlet, 97
nasal drug delivery, 13, 90
  atomisation, 90
  insertion angle, 101
  lung deposition, 93
  nanoparticle, 92
  nozzle diameter, 101
  nozzle tip, 101
  spray cone angle, 101, 103
Navier–Stokes, 83, 117
Newtonian fluid, 83

organisation chart, 135

particle image velocimetry (PIV), 4, 90
particle size distribution, 110
particle trajectories, 99
particle velocity, 99
patent, 133
polymerisation, 30, 34
  photopolymer, 31
project life cycle, 142
project management
  auxiliary team members, 146
  constraints, 140
  functional team leader, 146
  key performance indicator (KPI), 145
  organisation structure, 143
  PMBOK Guide, 139
  project manager, 145
  technical leader, 146
project schedule planning, 142

radiofrequency, 13
rapid prototyping, 30
  phantom, 32
region of interest, 42
renal artery ostia, 43, 47, 58, 72, 76, 82, 84
respiratory
  pleural cavity, 97

scientific management, 127
scintillation counting, 90
segmentation, 79, 93
selective laser melting, 32
shear stress, 64

standard operation procedure, 134
stent graft, 11, 40, 70
    abdominal, 64
    aorta, 47, 72
    balloon-expanding, 70
    self-expanding, 70
    thoracic, 64
stereolithography (STL), 26, 30, 80, 93
suprarenal stent graft, 58, 78
surface reconstruction, 18, 25, 93
swirl fraction, 98, 101

trade secret, 133

ultraviolet laser, 30

virtual bronchoscopy, 42
virtual intravascular endoscopy (VIE),
        26, 40, 41, 58
    threshold, 43
visualisation, 4, 17
volume-averaged flow velocity, 124

Reviews for "Methods of Development and Research of Biomedical Devices"

"Clinicians, surgeons, researchers and entrepreneurs in the medical industry will find this book as a practical guide to developing new biomedical devices. This work is based on a more practical approach for describing the process of device development, and offers a blueprint on how to implement prototyping, testing, evaluation and delivery of a successful medical product. Congratulations to the authors for this timely contribution."

– Prof. Derek Abbott, University of Adelaide

"Medical devices play an increasingly important role in contemporary health care. This publication will contribute significantly to the advancement of innovation of medical devices through multidisciplinary collaboration and effective application of technologies for effective and efficient health care delivery."

– Prof. Charlie Xue, RMIT University

"This monograph is a statement on the current state of the art integration of multidisciplinary technologies in medical imaging and visualisation, high fidelity computational engineering simulations (which can also serve as a platform for enhancing a rational assessment for disease diagnosis), product design and manufacturing for the design and manufacture of biomedical devices."

– Prof. Murali Damodaran, IIT Gandhinagar